T0201330

FUNDAMENTALS OF PUBLIC SAFETY NETWORKS AND CRITICAL COMMUNICATIONS SYSTEMS

IEEE Press
445 Hoes Lane
Piscataway, NJ 08854

IEEE Press Editorial Board
Ekram Hossain, *Editor in Chief*

Giancarlo Fortino	Andreas Molisch	Linda Shafer
David Alan Grier	Saeid Nahavandi	Mohammad Shahidehpour
Donald Heirman	Ray Perez	Sarah Spurgeon
Xiaoou Li	Jeffrey Reed	Ahmet Murat Tekalp

IEEE Press Series on Network and Service Management
Veli Sahin and Mehmet Ulema, *Series Editors*

Advisory Board

Thomas Plevyak	Prosper Chemouil	Rolf Stadler
(Founding Editor)	Alexander Clemm	José M. S. Nogueira
Raouf Boutaba	James Won-Ki Hong	

IEEE Press Series on Network and Service Management provides technical reference books and textbooks on network and service management including management of specific technologies in the field of Information and Communications Technologies (ICT). The focus is on FAB, i.e. Fulfillment (Configuration/Security), Assurance (Fault/Performance), and Billing (Accounting). Books are intended for ICT professionals as well as other related specialists in the private sector, government, research, academia and professional societies around the world.

FUNDAMENTALS OF PUBLIC SAFETY NETWORKS AND CRITICAL COMMUNICATIONS SYSTEMS

Technologies, Deployment, and Management

Mehmet Ulema

Manhattan College, Riverdale, New York, USA

**IEEE Press
Series on
Networks and
Services Management**

Dr. Veli Sahin and
Dr. Mehmet Ulema, *Series Editors*

Copyright © 2019 by The Institute of Electrical and Electronics Engineers, Inc. All rights reserved.

Published by John Wiley & Sons, Inc., Hoboken, New Jersey.
Published simultaneously in Canada.

No part of this publication may be reproduced, stored in a retrieval system, or transmitted in any form or by any means, electronic, mechanical, photocopying, recording, scanning, or otherwise, except as permitted under Section 107 or 108 of the 1976 United States Copyright Act, without either the prior written permission of the Publisher, or authorization through payment of the appropriate per-copy fee to the Copyright Clearance Center, Inc., 222 Rosewood Drive, Danvers, MA 01923, (978) 750-8400, fax (978) 750-4470, or on the web at www.copyright.com. Requests to the Publisher for permission should be addressed to the Permissions Department, John Wiley & Sons, Inc., 111 River Street, Hoboken, NJ 07030, (201) 748-6011, fax (201) 748-6008, or online at http://www.wiley.com/go/permission.

Limit of Liability/Disclaimer of Warranty: While the publisher and author have used their best efforts in preparing this book, they make no representations or warranties with respect to the accuracy or completeness of the contents of this book and specifically disclaim any implied warranties of merchantability or fitness for a particular purpose. No warranty may be created or extended by sales representatives or written sales materials. The advice and strategies contained herein may not be suitable for your situation. You should consult with a professional where appropriate. Neither the publisher nor author shall be liable for any loss of profit or any other commercial damages, including but not limited to special, incidental, consequential, or other damages.

For general information on our other products and services or for technical support, please contact our Customer Care Department within the United States at (800) 762-2974, outside the United States at (317) 572-3993 or fax (317) 572-4002.

Wiley also publishes its books in a variety of electronic formats. Some content that appears in print may not be available in electronic formats. For more information about Wiley products, visit our web site at www.wiley.com.

Library of Congress Cataloging-in-Publication Data is available.

hardback: 9781119369479

Cover design: Wiley
Cover image: © MarsYu/Getty Images

Printed in the United States of America.

V10006066_111618

To all those who lost their lives for keeping the public safe!

CONTENTS

FOREWORD BY ALAN KAPLAN

I have known Dr. Ulema for more than a decade. We have collaborated on several professional activities, including conferences and publications. Most recently, we served as co-guest editors for a special issue on the very same topic in IEEE Communications Magazine. Dr. Ulema is a leading expert in telecommunications, combining his experience in the industry with academic and scholarly work.

The field of public safety networks and critical communications has become enormously vital in the face of ever-increasing disasters and terrorist activities throughout the world. Communications and information technologies used by public safety agencies are being upgraded to provide much higher bandwidth and superior performance to support the multimedia applications demanded by these agencies. However, this is not a simple upgrade. It requires new protocols, new applications, new policies, etc., which may take a long time to realize. This means that existing technologies will be around for a long time, perhaps side by side with emerging technologies. Therefore, issues such as interoperability, migration planning, operations, and spectrum sharing need to be addressed.

Dr. Ulema's book provides extensive coverage of significant technologies, namely, P25, TETRA, DMR, and LTE, as well as potential advanced technologies and research topics. Spectrum, policies, and economics-related topics are discussed in detail. Extensive material about systems, services, end-user devices, and applications, along with planning, designing, and deployment-related topics, as well as network management aspects for sustainable operations, is provided.

The book is a must-have reference material filled with a wealth of information for professionals working in the field as well as academics researching various aspects of this critically important area.

Dr. Alan Kaplan
Lecturer, Princeton University, New Jersey
Chief Innovation Officer, Drakontas, LLC, Glenside, Pennsylvania

FOREWORD BY
HUSSEIN MOUFTAH

Ever since the 9/11/2001 attack destroyed the twin towers in New York City and Katrina devastated New Orleans in 2005, public safety networks, especially their use in coordinating the activities of many agencies involved in emergency operations, have gained visibility and raised many questions about their effectiveness. Many agencies still use century-old analog systems, while the use of digital narrowband technologies, such as P25 and TETRA, has become more prevalent. Further, there are a growing number of efforts to introduce broadband-based technologies that are to be used in public safety networks. Unfortunately, there are a relatively small number of publications and scholarly books on this important field.

Dr. Ulema is a leading expert in the field of telecommunication, with extensive experience in the telecom industry as well as in academia. In this book, he integrates insights regarding communications and networking characteristics, their technical and economic feasibility, design, deployment, as well as management for sustainable operation of such networks.

His book is a treasure for professionals working in the field as well as researchers working toward advancements in the discipline. Anyone interested in this field will be delighted by the comprehensive contents of this book.

Dr. Hussein T. Mouftah
Distinguished University Professor and Tier 1 Canada Research Chair
School of Electrical Engineering and Computer Science
University of Ottawa, Canada

PREFACE

This book is about public safety networks and mission-critical communications systems. The objective is to provide a comprehensive sourcebook covering the communications technologies that may be used in public safety networks and mission-critical communications systems. Also, the book covers a number of closely related areas such as the design, deployment, management, and operation of such networks.

Public safety networks have always been critically important for public safety agencies such as police departments, firefighters, and ambulatory services, especially in dealing with emergency situations such as natural and manmade disasters. Mission/business critical communications networks have been used by many commercial and non-commercial organizations in various sectors such as construction, transportation, factories, and mining operations as well.

The underlying networking technology has a significant influence on the characteristics of end-user devices, application supporting systems, operations support systems, and operation and management of networks. Surprisingly, old analog radio technologies are still in use in significant numbers today. However, narrowband digital radio technologies such as Project 25, Terrestrial Trunked Radio (TETRA), and Digital Mobile Radio (DMR) are gradually replacing analog technologies. Furthermore, many agencies and countries have started to entertain the idea of using broadband technologies for their public safety networks. Long-Term Evolution (LTE) technology and its iterations are the strongest candidates for this because of their success, partly due to the availability of commercial broadband applications.

Planning, designing, and deployment of public safety networks and mission-critical communication systems depend on many factors, including the type of organization, the number of organizations sharing the system, coverage, interoperability, existing systems, data requirements, nationwide plan, finance, and frequency spectrum.

The book covers many of the areas discussed above in four distinct parts.

The first part includes three chapters, which provide a detailed introduction. Chapter 1 provides an overview of the field. It is a comprehensive summary of the book. It can be considered an executive overview. Chapter 2 identifies the users of critical communications systems and public safety networks. Finally, Chapter 3 discusses the characteristics of critical communications systems.

The second part deals with communications technologies, which are covered in six chapters. Chapter 4 first provides an overview of the technologies and standards available for critical communications and public safety networks. Chapters 5, 6, and 7 present narrowband technologies—Project 25, TETRA, and DMR, respectively. Chapter 8 is dedicated to LTE technology, focusing on critical communications related features. The last chapter in this part, Chapter 9, focuses on the emerging technologies that are likely to be used for critical communications systems.

The third part includes two chapters providing a discussion on the applications, systems, and end-user devices that are traditionally used in public safety networks and critical communications systems. Chapter 10 focuses on support systems and applications and Chapter 11 discusses end-user equipment.

The fourth part is dedicated to the planning, deployment, and management of public safety networks and critical communication systems. Chapter 12 discusses planning for deploying and operating critical communications systems. Chapter 13 addresses economic and financial considerations. Chapter 14 focuses on design and deployment issues. Finally, Chapter 15 provides a detailed discussion on the operation, administration, and management of public safety networks.

The last chapter of the book, Chapter 16, provides a summary and several conclusions. The book also provides a complete list of standards documents in three appendixes: Appendix A contains a list of Project 25 standards documents, Appendix B provides a list of TETRA documents, and Appendix C provides a list of LTE standards documents related to public safety applications.

We believe that this book is of immediate interest to professionals in the industry (operators, manufacturers, application developers, system integrators, testers, etc.), government agencies, and regulatory agencies that are dealing with or interested in the development of public safety networks and critical communications systems. Also, this book will be useful to researchers in academia and the industry who are investigating various aspects of critical communications systems and public safety networks. Furthermore, we anticipate and hope that these discussions encourage further research and development, leading to more advanced solutions.

Also, the book will be useful to decision makers in the government and industry as they consider the migration of services to next-generation broadband-based critical communications networks. As a contemporary and comprehensive book, it will also attract the growing global community of professionals associated with this field as well as others who wish to gain perspective.

The book may also be used as a textbook for a graduate course in telecommunications. Chapters 1 through 9 may be used for a course focusing on the technology. Chapters 1, 4, and 9 through 15 may be used in a graduate course focusing on the design deployment and management of such networks.

There are just a few books dedicated to this vital topic. Some of them are edited books; different authors write each chapter. Some books focus on public safety networks, especially broadband technology-based ones. No significant discussion has been provided in the literature on the critical communications systems that are used by commercial organizations. Some other books are dedicated to LTE-based public safety networks only. These books do not cover the full spectrum, from academics, research, and business to the practice. Therefore, as of this writing, to the best of our knowledge, there is no comprehensive book covering all aspects of the public safety and critical communications network field together. This book addresses this void by creating a comprehensive book for the broad audience described above.

MEHMET ULEMA

ACKNOWLEDGMENTS

While I was working on a project related to public safety networks, I had a difficult time finding comprehensive and reliable literature. Yes, there were pieces here and there, but I had to spend a significant amount of time and effort to gather and assimilate them. I saw that there was a void and decided to write this book.

I have been lucky to have encouragement, support, and help provided by many colleagues and friends from the beginning, all the way to the end of this endeavor.

I would like to acknowledge with gratitude the support I received from the School of Business at Manhattan College, and especially Dean Salwa Ammar for her encouragement and support. I want to thank my student Zerena Lupo for assembling the list of standards documents related to the technologies used in public safety networks.

Special thanks go to my good friend and colleague, Professor Kudret Topyan, who taught me everything I know about finance and economics. He was gracious enough to review the chapter on economics and finance and helped me to put together the final version. I also want to thank Barcin Kozbe, who helped me to gather material on the operation and management of these networks, and Hakki Candan Cankaya, who reviewed several chapters and provided valuable feedback.

I am most grateful to Dr. Hussein Mouftah and Dr. Alan Kaplan, who graciously provided generous forewords for the book.

I am especially indebted to Dr. Veli Sahin, co-editor of the IEEE Press Book series on Network and Services Management. He and his co-editor, Tom Plevyak, guided me throughout the process, reviewed the manuscripts, and provided valuable comments.

I am grateful to all of those in IEEE Press and Wiley with whom I have had the pleasure to work with during this project.

Finally, this book would not have been possible without the support and encouragement of my wife, Terrie Ulema. For indulging my long nights at the computer, I'd like to thank my wife and my kids, Peri and Deniz. They all kept me going.

LIST OF ABBREVIATIONS

2G	2nd Generation
3G	3rd Generation
3GPP	3rd Generation Partnership Project
4G	4th Generation
a.k.a.	also known as
ACELP	Algebraic Code Excited Linear Prediction
AES	Advanced Encryption Standard
AFSI	Analog Fixed Station Interface
AI	Air Interface
AI	Artificial Intelligence
AIS	Application Interface Specification
AKA	Authentication and Key Agreement
ALI	Automatic Location Identification
AM	Amplitude Modulation
AMBE	Advanced Multi Band Excitation
AMR	Adaptive Multiple Rate
ANI	Automatic Number Identification
ANSI	American National Standards Institute
AP	Application Protocol
APCO	Association of Public Safety COmmunications
AR	Augmented Reality
ARC	American Red Cross
AT	Attention
ATIS	Alliance for Telecommunications Industry Solutions
AVL	Automatic Vehicle Location
b/s/H	bits per second per Hertz
BER	Bit Error Rate
BML	Business Management Layer
bps	bits per second
BS	Base Station
BSC	Base Station Controller
BMS	Business Management System
BT	Relative filter bandwidth

BYOD	Bring Your Own Device
C4	Computing, Command, Control, and direction-finding Communications
C4FM	Continuous 4 level FM
CAI	Common Air Interface
CAP	Common Alerting Protocol
CAP	Compliance Assessment Program
CAPEX	CAPital EXpenses
CBC	Cell Broadcast Center
CBS	Cell Broadcast Service
CCBG	Critical Communications Broadband Group
CCC	Composite Control Channel
CDMA	Code division multiple access
CMCE	Circuit Mode Control Entity
CMIP	Common Management Information Protocol
CoMP	Coordinated Multi-Point transmission and reception
CQPSK	Compatible Quadrature Phase Shift Keying
CR	Cognitive Radio
CRV	Call Retention Value
CSFB	Circuit-Switched FallBack
CSS	Console Sub System
CSSI	Console Sub System Interface
CTCSS	Continuous Tone-Controlled Squelch System
CITIG	Canadian Interoperability Technology Interest Group
D2D	Device to Device
DCC	Dedicated Control Channel
DCS	Digital-Coded Squelch
DeNB	Donor eNB
DES	Data Encryption Standard
DFSI	Digital Fixed Station Interface
DGNA	Dynamic Group Number Assignment
DHS	Department of Homeland Security
DIMRS	Digital Integrated Mobile Radio Service
DL	Down Link
DLL	Data Link Layer
DMO	Direct Mode Operations
DMR	Digital Mobile Radio
DNI	Data Network Interface
DQPSK	Differential Quadrature Phase Shift Keying
DSS1	Digital Subscriber System No. 1
DtD	Device to Device
EADS	European Defense & Space
EAS	Emergency Alert System

EDGE	Enhanced Data rates for GSM Evolution
eICIC	enhanced Inter-Cell Interference Coordination
eIMTA	enhanced Interference Management and Traffic Adaptation
EIR	Equipment Identity Register
eMBMS	enhanced Multimedia Broadcast Multicast Services
EMG	Emniyet Genel Müdürlüğü
EML	Element Management Layer
EMS	Element Management System
EMS	Emergency Management System
eNB	Enhanced Node Base station
eNodeB	Enhanced Node Base station
EPC	Evolved Packet Core
EPG	Exterior Gateway Protocols
ERC	European Radio communications Committee
eTOM	enhanced Telecommunications Operations Map
ETS	European Telecommunication Standard
ETSI	European Telecommunications Standards Institute
EU	European Union
E-UTRAN	Evolved Universal Terrestrial Radio Access Network
FCC	Federal Communications Commission
FDD	Frequency Division Duplex
FDMA	Frequency Division Multiple Access
FHMA	Frequency Hopping Multiple Access System
FirstNet	First Responder Network Authority
FM	Frequency Modulation
FMSS	Flexible Mobile Service Steering
FNE	Fixed Network Equipment
FPIC	Federal Partnership for Interoperable Communication
FSI	Fixed Station Interface
Gbps	Giga bits per second
GCSE	Group Call System Enablers
GERAN	GSM/EDGE Radio Access Network
GERYON	Next Generation Technology Independent Interoperability of Emergency Services
GHz	Giga Hertz
GIS	Geographic Information Systems
GMSK	Gaussian Minimum B14Shift Keying
GoTa	Global Open Trunking Architecture
GPRS	General Packet Radio Service
GPS	Global Positioning System
GSM	Global System for Mobile Communications
GTP-U	GPRS Tunneling Protocol-User Plane
GW	Gateway

HA	High Availability
H-CPM	Harmonized Continuous Phase Modulation
HD	High Definition
H-DQPSK	Harmonized Differential Quadrature Phase Shift Keying
HeNB	Home eNB
HeNB-GW	Home eNB GW
HetNet	Heterogeneous Network
HMD	Head-Mounted Display
HPUE	High-Power User Equipment
HSDPA	High Speed Downlink Packet Access
HSPA	High Speed Packet Access
HSS	Home Subscriber Server
HSUPA	High Speed Uplink Packet Access
HUD	Head-Up Display
IAAS	Infrastructure-As-A-Service
IARU	International Amateur Radio Union
ICIC	Inter-Cell Interference Coordination
ICT	Information and Communication Technology
ICTA	Information and Communication Technology Authority
ID	IDentification
IDRA	Integrated Dispatch Radio
IEC	International Electrotechnical Commission
IEEE	Institute of Electrical and Electronics Engineers
IKI	Inter Key Management Facility Interface
IMBE	Improved Multi-Band Excitation
IMS	IP Multimedia Subsystem
IMT	International Mobile Telecommunications
IoE	Internet of Everything
IoT	Internet of Things
IP	Internet Protocol
IPAWS	Integrated Public Alert and Warning System
IPG	Interior Gateway Protocols
IPI	IP Inter-working interface
IPSec	Internet Protocol Security
IPv4	IP version 4
IPv6	IP version 6
IRR	Internal Rate of Return
IRS	Internal Revenue Service
ISDN	Integrated Services Digital Data Network
ISO	International Organization for Standardization
ISSI	Inter-Sub System Interface
IT	Information Technology
ITIL	Information Technology Infrastructure Library

ITU	International Telecommunication Union
ITU-R	International Telecommunications Union - Radio communication Sector
ITU-T	International Telecommunications Union - Telecommunications Sector
Kbps	Kilo bits per second
KDI	Key Fill Device Interface
KFD	Key Fill Device
kHz	kilo Hertz
km	kilo meter
km/h	km per hour
KMF	Key Management Facility
KPI	Key Performance Indicators
L1	Layer 1
L2	Layer 2
LAN	Local Area Network
LCS	Location Services
LDT	Line Dispatch Terminal
LDU	Logical Data Unit
LIP	Location Information Protocol
LLC	Logical Link Control
LMR	Land Mobile Radio
LS	Line Station
LTE	Long Term Evolution
LTE-A	LTE-Advanced
LTE-D	LTE Direct
M2M	Machine to Machine
MAC	Medium Access Control
Mbps	Megabits per second
MBSP	Multimedia Broadcast Supplement for Public Warning System
MC	Mission Critical
MC-PTT	Mission-Critical PTT
MDP	Mobile Data Peripherals
MEF	Metro Ethernet Forum
MELPe	Mixed Excitation Liner Predictive, enhanced
MESA	Mobility for Emergency and Safety Applications
MHz	Mega Hertz
MIMO	Multiple-Input/Multiple-Output
MIMS	Minnesota Incident Management System
MINSEF	Minnesota Statewide Emergency Frequency
MIRR	Modified Internal Rate of Return
MME	Mobility Management Entity
MoM	Manager of Managers

MOP	Method Of Procedures
MoU	Memorandum of Understanding
MPLS	Multiprotocol Label Switching
MPT	Ministry of Posts and Telegraph
ms	milli second
MS	Mobile station
MSAG	Master Street Address Guide
MSC	Mobile Switching Center
MSO	Mobile Switching Office
MVNO	Mobile Virtual Network Operator
NAS	Non-Access Stratum
NASTD	National Association of State Telecommunications Directors
NATO	North Atlantic Treaty Organization
NCS	National Communications System
NCT	New Carrier Type
NEL	Network Element Layer
NFV	Network Functions Virtualization
NG911	Next Generation 911
NGO	NonGovernmental Organization
NID	Network Interface Devices
NIMS	National Incident Management System
NIST	National Institute of Standards and Technology
NMC	Network Management Center
NMI	Network Management Interface
NML	Network Management Layer
NMS	Network Management System
NOAA	National Oceanic and Atmospheric Administration
NOC	Network Operations Center
NOCC	Network Operations Control Center
NPR	National Public Radio
NPSTC	National Public Safety Telecommunications Council
NPV	Net Present Value
NR	New Radio
NSA	National Security Agency
NSN	Nokia Siemens Networks
NTIA	National Telecommunications and Information Administration
NVT	Network Validation Test
NXDN	Next Generation Digital Narrowband
OA&M	Operations, Administration and Maintenance
OFDM	Orthogonal Frequency Division Multiplexing
OFDMA	Orthogonal Frequency Division Multiple Access
OMA	Open Mobile Alliance
OMC	Operations and Maintenance Center

OPEX	OPerational EXpenses
ORT	Operation Readiness Test
OSS	Operations Support Systems
OTAP	Over The Air Programming
OTAR	Over The Air R+B47ekeying
OTIP	Over-The-Intranet-Programming
P25	Project 25
P25 PTToLTE	P25 PTT over LTE
PAMR	Public Access Mobile Radio
PAN	Personal Area Network
PAS	Publicly Available Specifications
PBX	Private Branch Exchange
PCM	Pulse-Code Modulation
PCRF	Policy and Charging Rules Function
PDCP	Packet Data Convergence Protocol
PDH	Plesiochronous Digital Hierarchy
PDN	Packet Data Network
PDP	Packet Data Protocol
PDS	Packet Data Services
P-GW	Packet Data Network Gateway
PM	Phase Modulation
PMR	Private Mobile Radio
PoC	Push-to-talk over Cellular
POP	Point of Presence
PPDR	Public Protection and Disaster Relief
PPP	Point-to-Point Protocol
ProSe	Proximity Services
PSAN	Public Safety Application Network
PSAP	Public-Safety Answering Point
PSCR	Public Safety Communications Research
PSRA	Public Safety Related Applications
PSS1	Private Signalling System 1
PSTN	Public Switched Telephone Network
PTIG	Project 25 Technology Interest Group
PTT	Push-To-Talk
PTToLTE	PTT over LTE
QAM	Quadrature Amplitude Modulation
QoE	Quality of Experience
QoS	Quality of Service
QPSK	Quadrature Phase Shift Keying
QSIG	Q SIGnaling
R&D	Research and Development
RAN	Radio Access Network

RCM	Radio Control Manager
RF	Radio frequency
RFI	Request for Information
RFID	Radio-Frequency IDentification
RFP	Request for Proposal
RFSS	RF Sub-System
RJ45	Registered Jack - 45
RLC	Radio Link Control
RN	Relay Node
ROI	Return ON Investment
RRC	Radio Resource Control
RRM	Radio Resource Manager
RTCP	Real-time Transport Control Protocol
RTP	Real-time Transport Protocol
S1-AP	S1 Application Protocol
SA	Study Area
SA	Service Availability
SA6	System Architecture 6
SAAS	Software-As-A-Service
SAT	Site Acceptance Testing
SBS	Site Base Station
SC-FDMA	Single Carrier FDMA
SCN	Switching Control Node
SCTP	Stream Control Transmission Protocol
SDH	Synchronous Digital Hierarchy
SDN	Software-Defined Networking
SDO	Standards Development Organization
SDR	Software Defined Radio
SDS	Short Data Services
SEC	Security Exchange Commission
SEG	Security Gateway
SFPG	Security and Fraud Prevention Group
SGSN	Serving GPRS Support Node
S-GW	Serving Gateway
SIP	Session Initiation Protocol
SISO	Single Input Single Output
SLA	Service Level Agreement
SML	Service Management Layer
SMS	Service Management System
SMS	Short Messaging Service
SNMP	Simple Network Management Protocol
SON	Self-Organizing Network
SONET	Synchronous Optical Network

SOP	Standard Operating Procedure
SoR	Statement of Requirements
SP	Service Provider
SUV	Sport-Utility Vehicle
SRVCC	Single Radio Voice Call Continuity
SVLTE	Simultaneous voice and LTE
SwMI	Switching and Management Infrastructure
TBO	Total Benefits of Ownership
TCCA	TETRA Critical Communications Association
TCO	Total Cost of Ownership
TCP	Transmission Control Protocol
TDD	Time Division Duplex
TDMA	Time Division Multiple Access
TEA	TETRA Encryption Algorithm
TEDS	TETRA Enhanced Data Service
TEI	Terminal Equipment Interface
TETRA	Terrestrial Trunked Radio
TF	Task Force
TIA	Telecommunications Industry Association
TII	Telephone Interconnect Interface
TM	Telecommunications Management
TMF	Telecommunications Management Forum
TMN	Telecommunications Management Network
TMO	Trunked Mode Operation
TMS	Terminal Management System
TNP1	TETRA Network Protocol 1
TR	Technical Report
TS	Technical Standard
TSBK	Trunking Signaling Block
TV	Television
UAV	Unmanned Aerial Vehicle
UCM	User Configuration Manager
UDP	User Datagram Protocol
UE	User Equipment
UHF	Ultra High Frequency
UICDS	Unified Incident Command and Decision Support
UK	United Kingdom
UL	Up Link
UMTS	Universal Mobile Telecommunications System
UNC	Unified Network Configurator
UO	Up Link
US	United States
USA	United States of America

USB	Universal Serial Bus
USCG	United States Cost Guard
USD	US Dollar
UTRAN	Universal Terrestrial Radio Access Network
V2V	Vehicle-to-Vehicle
V2X	Vehicle-to-Everything
VHF	Very High Frequency
ViLTE	Video over LTE
VLAN	Virtual LAN
VLSI	Very Large Scale Integration
VoIP	Voice over Internet Protocol
VoLTE	Voice over LTE
VR	Virtual Reality
WACC	Weighted Average Cost Of Capital
WCDMA	Wideband Code Division Multiple Access
WDM	Wavelength Division Multiplexing
WEA	Wireless Emergency Alerts
Wi-Fi	Wireless Fidelity
WiMAX	Worldwide Interoperability for Microwave Access

ABOUT THE AUTHOR

Dr. Mehmet Ulema is a professor of Computer Information Systems at Manhattan College, New York. Previously, he held management and technical positions in AT&T Bell Laboratories, Bellcore, Daewoo Telecom, and Hazeltine Corporation. Dr. Ulema's more than 30 years of experience in telecommunications as a professor, researcher, systems engineer, project manager, network architect, and software developer can be summarized as follows:

- Research, teaching, and consulting in information and communications technology areas, including critical communications and public safety networks, software-defined networks, wireless mobile networks, and network management
- Management of R&D teams in wireless mobile networks, intelligent wireless networks, network management, router development, and protocol development
- Design and development of algorithms and protocols in wireless networking, intelligent networking, data communications, and network management
- Extensive experience in national/international telecom standardization processes in leadership positions.

Dr. Mehmet Ulema has been actively involved in many major international conferences. He served as the General Chair, Technical Program Chair, and other leadership positions. Dr. Ulema is the recipient of the 2015 IEEE Communications Society Harold Sobol Award for Exemplary Service to Meetings & Conferences.

He has authored numerous papers, book chapters, and organized many special issues in several journals and technical magazines. He has been on the editorial board of a number of journals, including the IEEE Journal of Internet of Things, IEEE Transactions on Network and Service Management, and the Springer Journal of Network and Services Management. In the past, he served on the editorial boards of several other journals, including Elsevier Journal of Computer Networks and ACM/Springer Journal of Wireless Network.

Dr. Ulema received MS & Ph.D. in Computer Science at Polytechnic University (now the NYU Tandon School of Engineering), Brooklyn, New York, USA. He also received BS & MS degrees at Istanbul Technical University, Turkey.

1

OVERVIEW

This chapter provides an overview of public safety networks and critical communications systems. It is intended to be an executive summary. To provide a complete *picture*, some of the material (figures, text, etc) in other chapters are repeated here.

1.1 BACKGROUND

This book is a comprehensive treatment of technologies and systems used and to be used in public safety networks and mission-critical communications systems. The book also covers economic, financial, and policy issues as well as the design, deployment, and operation of such networks. Before we go further, let's explain what we mean by these networks:

A **public safety network** is a communications system used by the agencies involved in public safety affairs. The communications system used by a police department is an example of a public safety network. Typical functions include first and emergency

Fundamentals of Public Safety Networks and Critical Communications Systems: Technologies, Deployment, and Management,
First Edition. Mehmet Ulema.
© 2019 by The Institute of Electrical and Electronics Engineers, Inc. Published 2019 by John Wiley & Sons, Inc.

responses to wide-scale natural disasters such as earthquakes, forest fires, flooding, and man-made disasters such as nuclear explosions, radiation, terrorism, as well as localized emergencies such as automobile accidents, fires, medical emergencies, and any other threats to public order. (Note that in Europe, the term PPDR, short for Public Protection and Disaster Relief, is used to refer to public safety and first responders networks).

A **mission-critical communications system** is a network used by organizations to provide communications infrastructure to carry out mission-critical functions. The communications system used by workers at a large construction site is an example of a mission-critical communications network. Mission-critical communications networks have been used in various sectors, such as construction, transportation, utilities, factories, and mining operations. (Note that in some literature, the term "mission-critical communication" is used to refer to the communications systems used by law enforcement and emergency services as well [1]).

Although public safety networks and mission-critical networks differ in scale, design, deployment, and operations, the technologies used by both types of networks are highly similar. Therefore, in this book, we adopt the words "critical communications" to refer to both public safety and mission/business critical communications systems and networks. Occasionally, we may use these names interchangeably.

Critical communications systems include a telecommunications network with wireless and wired components, a set of services and applications, a variety of end-user devices, as well as some operations support systems, also known as network management systems. Critical communication systems also make use of radio frequency bands to exchange voice, data, and multimedia applications needed to carry out their "critical" functions as well as to transmit and receive information among users in the field and technicians at command centers.

Figure 1.1 gives an idea of market segmentation of the critical communications field for a specific narrowband technology, namely Terrestrial Trunked Radio (TETRA).

What sets a critical communications network apart from a commercial communications network? Perhaps the most dominant characteristics of critical communications networks are that they provide the basis for *situational awareness* and *command and control* capabilities, which roughly translate into the following capabilities [3]:

- prioritize delivery of mission-critical data (e.g. bring the dispatch data into the field: ability to send more and detailed information to the officers in real time),
- survive multiple failures (robust, even in extreme conditions; site hardening; enhanced physical protection and battery back-up; redundancy [intra-network, inter-network; fallback to other networks when needed]),
- maintain data integrity and confidentiality (end to end full encryption; link security—both user and control planes; network operations and management security including related data),

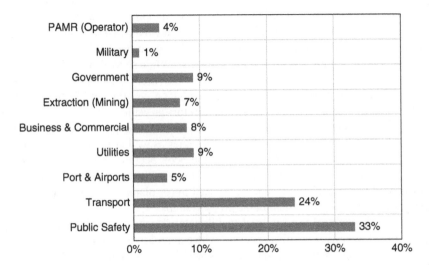

Figure 1.1. Sectors with TETRA-based critical communications [2].

- offer the essential coverage and capacity required (geographical coverage, not population coverage; symmetric usage [uplink-downlink] pattern, as opposed to downlink heavy commercial pattern),
- interoperate with other networks and extend coverage and capacity when needed (to enable communication among users outside network coverage, and to secure wide area communication even when users are outside normal network reach), and
- provide right to use and identity management support for officers, applications, and devices (to provision users with "right to use" of resources; dynamic priority and resource management for users and applications).

Many of the existing critical communications technologies have been in use for about 20 years now. They are mature, reliable, and relatively cost-effective in supporting critical voice applications. However, they are not designed to provide higher bandwidth supporting multimedia applications, which are requested by public safety agencies. Many countries have initiated projects to develop dedicated, nationwide public safety broadband networks to address these and other issues. For example, an authority called First Responders Network (FirstNet) was created in the USA in 2012 to establish such a national public safety broadband network.

Various industrial sectors having critical communication networks are also going through similar evolutionary phases. Superior capabilities and economy of scale offered by broadband technologies provide a rather attractive solution for upgrading the existing systems.

The standards organizations responsible for developing narrowband technologies has stated publicly that their future strategy is to be involved in developing Long Term Evolution (LTE)-based solutions for critical communications systems. However, it is expected that this transition will take a long time. While several countries have been planning for an LTE-based system, procurement activities for establishing nationwide systems based on older technologies such as TETRA and Project 25 are still taking place. The same trend is true in many other sectors as well.

Therefore, the primary objective of this book is to provide comprehensive coverage of the existing public safety technologies as well as the other technologies considered for future plans. We hope that the book becomes a valuable source for designing, deploying, and managing critical communications networks based on the narrowband and broadband technologies used in (or planned for) public safety networks and mission-critical communications systems.

Note that "national security" and "public safety" are two related, but separate topics. National security is mostly concerned with external/internal threats, whereas public safety concerns include natural disasters, accidents, and deliberately harmful acts.

1.2 TECHNOLOGIES USED IN CRITICAL COMMUNICATIONS

Old analog critical communications radio technologies have been replaced in most of the world by narrowband, all-digital, and voice and data technologies. Currently, narrowband digital radio systems are the primary technology used by public safety agencies and by many sectors. These systems are referred to as Land Mobile Radio (LMR) or Private Mobile Radio (PMR) systems, which are based on mainly Project 25, TETRA (and its variations), and Digital Mobile Radio (DMR) standards. TETRA has been the choice of public safety agencies and commercial and public organizations mainly in Europe and Project 25 technologies have been used mainly in North America. DMR-based systems, a newer narrowband, all-digital, standard technology, have also been chosen in some regions.

Partly due to the availability of commercial broadband applications, and partially due to increasing demand by public safety agencies, the possibilities of broadband data services for public safety networks are being discussed increasingly in many developed countries, including the USA and European countries. LTE technology is at the center of this new trend [4, 5].

1.2.1 Narrowband Land and Private Mobile Radio Systems

Project 25 is the code name for a technology based on the standards developed by the Telecommunications Industry Association (TIA) with the participation of the member organizations of the Association of Public Safety Communications Officials (APCO) and US federal agencies. More than 80 countries around the world have adopted

Project 25. Also, about 40 companies provide Project 25-compliant equipment and applications [6, 7].

TETRA is the code name for a technology based on the standards developed by the European Telecommunications Standards Institute (ETSI). TETRA is a trunked radio system, which became widely used in Europe first, then in many countries around the world. TETRA and Critical Communications Association (TCCA) estimates that more than 250 TETRA networks in more than 120 countries are deployed as of June 2016. TETRA uses TDMA technology with four user channels on one radio carrier. Packet data (low speed), as well as circuit data modes, are available. TETRA Enhanced Data Service (TEDS), included in TETRA 2, enables more data bandwidth to TETRA data service users. Although the standard is designed to provide up to 691 Kbps, in practice, users typically get a net throughput of around 100 Kbps. The low data rate is partially due to limitations in spectrum availability [8–10].

There is also another narrowband LMR technology, called TETRAPOL, which should not be confused with TETRA. TETRAPOL, not as popular as TETRA, is also a digital, cellular trunked radio system for voice and data communications with critical communications applications in mind. TETRAPOL was initially developed by a French company called Matra Communications. Today, TETRAPOL Forum leads the support and further development of TETRAPOL technology. TETRAPOL's air interface is based on FDMA and GMSK modulation; 12.5 kHz carrier spacing, along with 10 kHz carrier spacing, is available [11].

DMR is the code name for a technology based on another ETSI standard for PMR and used in Europe and several regions of the world as a low-cost entry-level radio system for commercial and public safety use. DMR offers a quick and cost effective replacement for analog systems with all the benefits of a digital solution. DMR provides voice, data, and some supplementary services [12–15].

Table 1.1 provides a comparison of significant features of Project 25, TETRA, and DMR technologies.

Among these three narrowband digital technologies, Project 25 and TETRA have been around for more than 20 years. Therefore, there is a mature, tested, interoperable

TABLE 1.1. A Comparison of Project 25, TETRA, and DMR Features

Functionality	P25 Phase 1	P25 Phase 2	TETRA	DMR
Standards Organization	TIA	TIA	ETSI	ETSI
Channel Access Method	FDMA	TDMA	TDMA	TDMA
Channel Bandwidth	12.5 kHz	6.25 kHz	25 kHz	12.5 kHz
Raw Data Rate	9.6 Kbps	9.6 Kbps	36 Kbps	9.6 Kbps
Number of Time Slots	N/A	2	4	2
Direct Mode	Yes	Yes	Yes (DMO)	Yes (Tier 1)
Repeater (Talk-Through) Mode	Yes	Yes	No	Yes (Tier 2
Trunking Mode	Yes	Yes	Yes (TMO)	Yes (Tier 3)
Analog Fallback	Yes	Yes	No	Yes

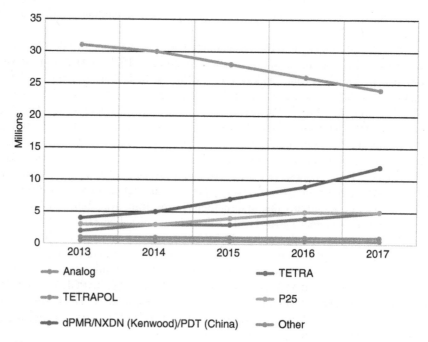

Figure 1.2. Global LMR subscriptions by technology: 2013–2017 (in millions) [16].

set of products available from many vendors. As shown in Figure 1.2, TETRA is the most widely used. Therefore, it is expected that equipment cost will be relatively lower than that of Project 25. DMR solutions may cost even less due to their less complicated architecture.

Project 25, TETRA, and DMR technologies are limited to providing data rates around 9.6–36 Kbps, which is rather slow to handle today's data-intensive applications, which require several megabits per second (Mbps) data rates. Therefore, public safety agencies have been looking into mobile broadband technologies to provide higher data rates [2, 17]. Lower indoor and rural handheld coverage and limited interoperability are some other design and economical limitations of these narrowband systems. ETSI and TIA agreed to work on a joint project called MESA to produce some specifications for a broadband standard for the critical communications ecosystem [18]. However, this was abandoned later on with the emergence of the concept of using LTE technology for critical communications systems.

1.2.2 Broadband Technologies for Critical Communications

Many public safety agencies around the world have been already using commercial broadband services (such as 4G and Wi-Fi) for data in conjunction with their

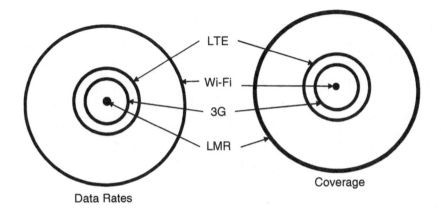

3G: Third Generation
Wi-Fi: Wireless Fidelity
LMR: Land Mobile Radio
LTE: Long Term Evolution

Figure 1.3. Coverage versus data rates [21].

voice-critical LMR systems. Furthermore, smartphones, tablets, and laptops have already been included as end-user devices by many agencies.

Although there are some different commercial mobile systems (such as Wi-Fi, WiMAX [19], and LTE) that are qualified as broadband technologies, there is a worldwide consensus that LTE is the technology choice for next generation critical communications systems [20–22] (see Figure 1.3 for a comparison of two important aspects of networking for some technologies). The US government recognized this and decided in 2009 on LTE as their platform for a national public safety network. Many other countries including China, England, Germany, Australia, and Qatar have also been focusing only on LTE-based public safety networks [16, 23–33]. Therefore, this book focuses on LTE-based broadband critical communications systems.

LTE is the only accepted technology worldwide as the fourth generation (4G) of mobile broadband communications systems. It is an evolution of second generation (2G), Global System Mobile (GSM), and third generation (3G) Wideband Code Division Multiple Access (W˙CDMA) technologies. LTE-Advanced (LTE-A), the next version of LTE, is the "true 4G" because unlike ordinary LTE, it meets the 4G system requirements (such as higher speed) set by the International Telecommunication Union (ITU) [4]. LTE-A provides better coverage, greater stability, and faster performance. LTE-Advanced supports up to 100 MHz bandwidth and 1–3 Gbps (downlink) peak data rate (note that these are theoretical numbers). Carrier aggregation, one of LTE-A's capabilities, allows operators to combine their separate narrow channels into one broader channel [34] (LTE delivers data using a contiguous block of frequencies up to 20 MHz wide). This feature results in significant performance gain

TABLE 1.2. A Comparison of LTE and LTE-A Features

	LTE	LTE-A
Transmission Bandwidth (MHz)	≤20	≤100
Peak Data Rate (DL/UL) (Mbps)	300 (low mobility 75 (high mobility)	1000 (low mobility) 500 (high mobility)
Latency (ms) User Plane	<6	<6
Latency (ms) Control Plane	50	50

LTE, Long Term Evolution (per Release 8); LTE-A, LTE-Advanced (per Release 10); DL, Down Link; UL, Up Link; MHz, MegaHertz; ms, millisecond.

since the bandwidth available to a mobile device is now much larger [5, 34, 35]. See Table 1.2 for a list of three significant features of LTE and LTE-A.

LTE is growing rapidly all around the world. Reference [36] reported that LTE reached 270 million subscriptions and that number is projected to increase to around 1.3 billion by 2018. As of June 2014, LTE was commercially available on 146 networks in 107 countries. Additionally, as of June 2014, LTE-Advanced was commercially deployed on nine networks in seven countries worldwide [36]. As of June 2016, the same reference [36] reports that there were 1.29 billion LTE subscribers worldwide. It also reports that as of June 2016, there were 118 LTE-Advanced networks in 54 countries. They all point to the fact that LTE has been grown rapidly and is being adopted worldwide.

Building a new LTE-based network requires many billions of dollars. The site acquisition and site deployment associated with the radio access network require heavy spending. It is very crucial to prepare an efficient site deployment strategy by considering how to maximize coverage with less number of sites. The use of existing infrastructure is highly crucial. Also, any revenue generation opportunity, such as partnerships with utility companies, and commercial carriers should be seriously considered.

However, there are a few issues related to using LTE for critical communications systems [37]. A major one is the support (or lack thereof) of mission critical voice and Direct Mode Operations (DMO), which allow first responders to communicate directly with others when the infrastructure is not available. Some commercial LTE networks are used mainly for data and multimedia applications. For voice applications, the existing, older technologies (e.g. 3G networks) are used. While Voice over LTE (VoLTE) has been standardized and several commercial deployments have taken place, it is still not widespread [5, 38, 39]. It may take some time for VoLTE and DMO to become widely implemented and deployed. An example of a mission-critical voice application is "group calling," which allows a large number of first responders to be included in the same conversation.

The "3rd Generation Partnership Project" (3GPP), the standards developing organization that has been developing LTE technology related specifications, has been actively involved in incorporating critical communications related features into upcoming LTE specifications, including device-to-device communications [40–45].

The future beyond LTE-A is highly promising as well. There has been a plethora of talks and activities to define the requirements of the fifth generation (5G) of mobile cellular technologies, which is envisioned to increase capacity and performance in order of magnitude compared to that of the current systems [46]. The current estimate is that 5G-based commercial networks will show up around 2020. Additionally, other technologies such as the Internet of Things (IoT), augmented reality, etc., may become a part of 5G and be commercially available. When and if these broadband-based technologies become available and commercially (read economically) viable, it is expected that critical communications systems will make use of these new technologies as well.

1.2.3 Interoperability

One of the weakest links in the current critical communications, especially in the public safety area, is interoperability [47, 48]. In many countries, there are no centralized common public safety networks that all agencies can use. It is most likely that different agencies use different communications technologies (interoperability problems may still be present due to differences in implementation, operation, and even jurisdiction; see Figure 1.4 for a comical depiction of the interoperability concern). Natural and manmade disasters have showed us that all agencies cooperating during such disasters must be able to communicate to be able to help the public. Therefore,

Figure 1.4. An example of interoperability solutions [49].

interoperability among all the networks (regardless of the technologies) used by all agencies is a paramount interest.

Currently, a makeshift arrangement is used for interoperability between two or more incompatible radio systems (e.g. "analog patching" between networks). Proprietary solutions also include interoperability via gateways, which use the same protocol for translating voice and data. This facilitates radios and protocols with different technologies to communicate.

The word interoperability is a loaded one. Its most comprehensive definition includes "governance, standard operating procedures, technology, training/exercises, and usage of interoperable communications" [22]. From the communications aspect, the word is used to mean that, for a given standard technology, the components built by different manufacturers work together. For example, an agency building a Project 25-based network acquires equipment from vendor X and vendor Y. The agency would want some guarantee that this equipment provided by two different vendors works when they are connected. This is typically verified by a set of *conformance* tests. All the technologies mentioned in this book have a set of well-defined procedures and standards to obtain *certificates* to prove the interworking of the equipment built by different vendors.

The same word, "interoperability," is also used to mean that different networks owned by different agencies work together. For example, an agent on network A should be able to communicate with another agent on network B. Network A and network B could be based on the same technology or each may be based on a different technology.

There are a bunch of interfaces and capabilities required for each technology to make it work with other technologies (this should include the networks based on analog technologies, which may be around for a while) (Table 1.3).

Also, applications, administration, operations, and security systems of each network should be configured to interoperate. Furthermore, public safety agencies may use a commercial landline and mobile network as well as Wi-Fi networks, especially during emergencies. Therefore, interoperability scenarios should also include these types of networks [21].

The Project 25-based system is already backward compatible with the existing DMR and other analog systems [22]. Furthermore, interoperability with commercial

TABLE 1.3. An Illustration of Possible Interoperability Scenarios

	Analog	P25	TETRA	DMR	LTE
Analog	x	x	x	x	
P25	x	x	x	x	x
TETRA	x	x	x	x	x
DMR	x	x	x	x	x
LTE		x	x	x	x

systems is also essential, especially during emergencies. Since the emergency call number system is one of the primary triggers for public safety activities, it is crucial that public safety networks be interoperable with emergency call centers as well.

There are several vendors offering solutions to provide complete interoperability with the LTE-based and Project 25-based system [50]. Since the Project 25 inter-systems interface is based on IP/TCP standards including Session Initiation Protocol (SIP) and Real-time Transport Protocol (RTP), which are also included in the LTE standards, the interoperability between these two should be relatively straightforward [51].

GERYON (Next Generation Technology Independent Interoperability of Emergency Services) was an EU R&D project. Its objective was to integrate the communication networks used by emergency—ambulances, fire brigades, civil protection teams—and safety (PMRs) management bodies with new generation telephone networks (4G, LTE). The project work plan defined a series of design and implementation work packages aimed at developing a non-commercial demonstrator prototype. At the end of the project, all its objectives were successfully fulfilled, resulting in a fully working IMS compatible ecosystem capable of providing PMR grade communications while paving the way for future professional LTE networks [52].

1.3 APPLICATIONS, SYSTEMS, AND END-USER DEVICES

The technologies discussed in the previous section are just one of the parts that make up the critical communications ecosystem. A complete ecosystem encompasses smart applications, supported by a set of comprehensive systems, purpose-built, intuitive devices, and comprehensive services as well. In other words, providing top levels of safety and efficiency to enable better decisions is about more than just better equipment and technology; it is about delivering new ways to connect users to information and each other. The critical communications ecosystem should deliver anywhere, anytime access to multimedia information with the priority, resiliency, and security that public safety agencies demand.

1.3.1 Applications and Systems

A modern critical communications ecosystem must be equipped with a variety of applications, from necessary push-to-talk to IP telephony to comprehensive multimedia voice and data applications. With narrowband technologies such as TETRA and Project 25, due to their low data rates, commercially available mobile devices such as smartphones and tablets are not available to critical communications systems users. However, to supplement the applications provided by narrowband technologies, commercial smartphones, and tablets connected to either a Wi-Fi or a commercial carrier, are frequently used by critical communication users. It is expected that with the introduction of LTE-based critical communications systems, smartphones and tablets, as well as a variety of other multimedia devices, be a part of applications and devices available to first responders and law enforcement agencies. Body cameras, license plate readers, fingerprint scanners, virtual maps, and digital building plans are just some of the applications that are expected to be a part of the critical communications ecosystem.

APCO International is the world's largest organization of public safety communications professionals. APCO International maintains a website that provides an inventory of applications, referred to as APCO International's online Application Community (AppComm) [53]. The site has a collection of applications related to public safety and emergency response. Some of these applications (e.g. neighborhood crime map) can be used by the general public as well. These applications are typically mobile apps that are intended for use on a smartphone or tablet.

Systems and applications deployed and used in critical communications systems should allow the users of such systems to submit and retrieve information by end-user devices, terminals, as efficiently as possible.

While most applications are deployed over the Internet and mobile networks, there are just a few data applications over TETRA and Project 25 due to the low data rate provided by these narrowband technologies. However, there have been some offerings by various vendors to ease the concern somewhat. Some of these applications can even be easily modified by the user thanks to the APIs provided by the vendors. It is expected that these applications extend information availability to a variety of end-user devices. Via the vendor provided APIs, users can develop their solutions in addition to traditional applications such as a database, forms, image handling, webmail, and others.

There are some applications currently available for various sectors such as law enforcement agencies, first responders, transport, and utilities. Table 1.4 shows some examples.

TABLE 1.4. A Few Sector-Specific Examples [54]

Police	• Vehicle, driver, license information inquiry
	• Transmission of missing person(s) images
	• Crime report and stop & search forms
	• Vehicle incident report lookup
Airport	• Missing passenger information look-up and submission
	• Incident report form look-up and submission
	• Fuel figure submission
	• Webmail access
Field Service	• Safety inspection report look-up and submission
	• Missing part information & photo download via Intranet
	• Fault report look-up

There are several shared centers and supporting associated systems to serve all the users in a coordinated way. Two of the most important ones are briefly discussed below:

- *Incident Management System*—provides a mechanism for all the users and agencies to work together "to prevent, protect against, respond to, recover from, and mitigate the effects of incidents, regardless of cause, size, location or complexity" [55]. Although each agency will have its control, command, and management centers, a unified incident control center provides smooth coordination and sharing of resources and capacities [55, 56].

  For example, in the USA, there is a new Department of Homeland Security (DHS) project called Unified Incident Command and Decision Support (UICDS), which will be used to share information for emergency operations. UICDS will be used to manage and share incident information across state and local lines, as well as with other federal agencies. Employing uniform standards, UICDS is intended to solve information interoperability problems, which have been a significant issue especially among public safety agencies [56].

- *Operations and Control Systems*—responsible for maintaining, administering, operating, and managing the whole network in a reliable, secure way. There may be agency-wide or region-wide centers and systems with similar responsibilities. All these systems should be connected, and activities need to be coordinated.

1.3.2 End-User Terminals and Consoles

Terminal devices used by public safety agencies strictly depend on the critical communications technology deployed by each agency. For example, the user devices for TETRA technology will be different from the user devices for Project 25 technology. Similarly, LTE-based critical communications devices will be drastically different from their narrowband counterparts, handling and displaying multimedia, just like the smartphones and tablets used commercially.

Regardless of the technology used, end-user devices may be categorized as mobile radios, portable radios, and consoles.

Mobile radios are installed in a motor vehicle such as cars and motorcycles (Figure 1.5). Since mobile radios are attached to the vehicle, they are bulkier, larger, and heavier than portable radios. Mobile radios have some advantages over portables: much better range, higher power output, and powered by the vehicle battery (no worry about battery life).

Portable radios are always carried (handheld) by the users. Therefore, they are relatively small and lightweight. As seen in Figure 1.6, a portable radio has a microphone and speaker. Like any other wireless portable device, it has a dipole

Figure 1.5. An example of a mobile radio [57].

antenna, powered by a rechargeable battery. The advantages mentioned for mobile radios become disadvantages for portable radios: smaller range, battery life, and low power output.

Dispatch consoles are systems, but since they are used to monitor/control end-user devices, we discuss them in the end-user devices section (Section 11.4). They are used to monitor and control multiple groups at a single physical position. The example shown in Figure 1.7 includes a microphone as well as the capability to select and unselect speakers. It provides EMERGENCY control.

New products come with enhanced functionality like built-in GPS, Wi-Fi, and Bluetooth interfaces, encryption support, and personal alarm buttons. A range of accessories, such as chargers and headsets, is also available. Tablets, smartphones, vehicular modems, and USB data cards are expected to be widely available once LTE-based critical communications systems are in place.

Figure 1.6. Examples of Project 25 portable radios.

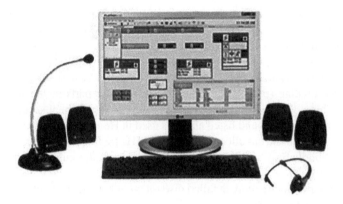

Figure 1.7. An example of a dispatch console [57].

1.4 STANDARDS, POLICIES, AND SPECTRUM

1.4.1 Frequency Spectrum for Critical Communications

Most voice land mobile radio systems in the USA use narrowband frequencies (12.5 kHz) in the VHF and UHF bands. However, the FCC has recently allocated 758–768 MHz and 788–798 MHz for base stations and mobile units use, respectively, 10 MHz wide for each direction for public safety applications. Also, for voice communications only, the 769–775 MHz and 799–805 MHz bands in 12.5 kHz narrowband increments are allocated for public safety use (Figure 1.8). The USA has also allocated a large band of the spectrum (50 MHz) in 4.940–4.990 GHz, although it is not clear how this would be used [58–60].

In Europe, the frequency bands 410–430 MHz, 870–876 MHz/915–921 MHz, 450–470 MHz, and 385–390 MHz/395–399.9 MHz are allocated for TETRA for civil use. For emergency services, the frequency bands 380–383 MHz and 390–393 MHz are allocated. If needed, these bands can be expanded from 383–395 MHz and

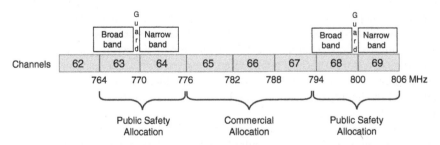

Figure 1.8. 700 MHz band plan for public safety services [61].

393–395 MHz to cover all of the spectrum [8, 62]. There is an ongoing effort in Europe to determine the most appropriate (and harmonized with other countries) frequency spectrum for broadband applications to be used by the public safety sector [8].

In several countries in Asia, like in Europe, the "380–400" MHz band is reserved for public safety organizations and the military as well. The 410–430 MHz band was allocated for civilian (private/commercial) use in other parts of the world, too. In Mainland China, the 350–370 MHz band is reserved for national security networks while the first 800 MHz band listed above is used in Hong Kong for private networks (Section 12.2.3). Russia has allocated 450–470 MHz for this purpose. In Australia and parts of the Middle East, spectrum has also been allocated for public safety broadband services

The book provides a more detailed discussion of how many countries around the world address spectrum issues.

1.4.2 Standards Development in Critical Communications

Traditionally, standardization of critical communications interfaces and protocols has been handled mainly in two Standards Development Organizations (SDOs), namely TIA for Project 25 and ETSI for TETRA and DMR-related projects. The standardization work on LTE-based standards has been carried out mainly by an entity called 3GPP, a collaboration among groups of telecommunications standards associations. It goes without saying that these SDOs do not operate in a vacuum. Many other organizations and even other SDOs provide input to this process. See Table 1.5 for a list of SDOs and other organizations involved in the standardization of critical communications systems.

In the following paragraphs, we briefly discuss several SDOs that play significant roles in the development of critical communications related specifications.

APCO and TIA collaborate, with some other organizations, to develop specifications for Project 25, also known as APCO Project 25, (which is the project name and number given by APCO) to produce public safety digital LMR standards. It is a joint project among the US APCO, the National Association of State Telecommunications Directors (NASTD), selected federal agencies and the National Communications System (NCS) in the USA, and the TIA. Project 25, designated as TIA-102, has been accepted as a national standard in the USA [6]. While APCO is the sole developer and formulator of the standard, TIA provides technical assistance and documentation for the standard. Project 25 is directed by a steering committee, which includes experts from various public safety agencies. Project 25 continues to evolve. The ongoing work in APCO, TIA, and other stakeholders has been centered on issues related to the interoperability between Project 25 and LTE-based public safety networks [63].

ETSI began the standards development for TETRA in the 1980s in Europe. The initial intent was to develop a standard for the wireless mobile network for commercial use. While ETSI was spending many years developing this comprehensive

TABLE 1.5. A List of SDOs and Other Organizations Involved in the Standardization of Critical Communications

SDOs, Organizations	Standards
APCO	P25
TIA	P25
ETSI	TETRA, TETRA/TEDS, DMR
3GPP	LTE, LTE-A
ATIS	All-IP and M2M infrastructure, Public Safety Related Applications Task Force (PSRA-TF), a Public-Safety Answering Point (PSAP)
ITU	Interoperability in Public Safety Mobile Networks, Spectrum for public safety communications
NPSTC (USA)	FirtsNet Requirements
TCCA	TETRA, TETRA to LTE Evolution
OMA	Push-to-talk over Cellular (PoC)

LTE, Long Term Evolution (per Release 8); NPSTC, National Public Safety Telecommunications Council; TCCA, TETRA and Critical Communications Association; TETRA, Terrestrial Trunked Radio; OMA, Open Mobile Alliance; APCO, Association of Public-Safety Communications Officials; TIA, Telecommunication Industry Association; ETSI, European Telecommunications Standards Institute; 3GPP, 3rd Generation Partnership Project; ATIS, Alliance for Telecommunications Industry Solutions; ITU, International Telecommunication Union.

system, GSM networks became popular and ubiquitous. This caused significant hardship for the companies invested in TETRA development and they identified the public safety market as a way to sell their products. TETRA in ETSI is expected to continue to provide enhancements only. In other words, ETSI has no plans to develop new technology in this area. The TETRA community has been active in moving toward LTE-based public safety networks as well. Some projects are underway to achieve seamless interoperability between TETRA and LTE-based public safety networks [64].

3GPP has been working on the standardization of LTE as part of the Release 8 feature set (a *release* in 3GPP refers to a group of added technology components). Following Release 8 and Release 9, significant improvements were incorporated into Release 10. With this new release, LTE got a new name as well: LTE-Advanced.

The first commercial LTE system was deployed in late 2009. 3GPP working groups added new features and technology components into later releases to improve LTE. Release 12 enhances LTE to meet public safety application requirements. Two critical public safety-related study items, Direct Mode Operation, and Group Call functions are included in Release 12 [43]. Public safety agencies and other stakeholders (such as TCCA, APCO, and FirstNet) are working together to drive the development of additional features that are typically associated with public safety systems [44, 65] (Figure 1.9).

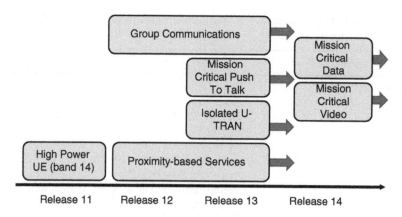

Figure 1.9. 3GPP critical communications related projects [66].

1.5 PLANNING, DESIGN, DEPLOYMENT, AND OPERATIONAL ASPECTS

Planning, designing, and deployment of a critical communication system depend on many factors, including the type of organization, the number of organizations to share the system, the coverage, interoperability, existing systems, data requirements, nationwide plan, finance, and frequency spectrum.

This section provides an overview of how planning, designing, and deployment of a critical communication system deal with these questions during various stages of making a new critical communications system a reality.

1.5.1 Planning

Although the level of effort, activities, and contents heavily depend on the factors mentioned above, there are some common activities in planning for a critical communications system. Common activities include conducting a set of feasibility studies, developing a business case, performing a risk analysis, drawing up a roadmap, developing a business plan and a project plan, and establishing a project team.

A **feasibility study** is an analysis of the viability of an idea, a project, through a disciplined and documented process of thinking [67]. It is perhaps the most critical phase in the development of a project [68]. The feasibility study is conducted before developing a formal business plan [69]. A feasibility study results in a feasibility report that provides documentation that the idea was thoroughly investigated. The feasibility report explains in detail whether the project under investigation should be carried out. A typical feasibility study includes operational feasibility, market feasibility, financial/economic feasibility, organizational/managerial feasibility, environmental feasibility, and legal feasibility. Note that not all these types may apply to a given feasibility study [69].

Before a feasibility study begins, we need to know what is being studied. Is it for upgrading an existing system or establishing a brand new system? Therefore, a set of criteria addressing the necessity, attainability, completeness, consistency, and complexity must be established as the first step in a feasibility study. These criteria will define conditions for the selected approach to be acceptable to the users and other stakeholders.

A set of more specific criteria that are directly related to the technology, operations, cost, finance, and the users must be established as well. Some examples are user expectations, data rate requirements, performance (throughput, capacity, and latency), availability, reliability, resiliency, security, scalability, evolvability, interoperability, manageability, cost-effectiveness, and more.

1.5.2 Technology Considerations for a Critical Communications System

Based on the discussions in Section 1.2, we can safely say that, currently, there are four major candidate technologies to consider: TETRA, Project 25, DMR, and LTE. Again, depending on what exists and what criteria and requirements are established, there may be different scenarios constructed. A table similar to Table 1.6 may be

TABLE 1.6. A Template for a Qualitative Comparison of Technology Considerations

		Criteria				
		Criterion 1	Criterion 2	Criterion 3	Criterion 4	Criterion 5
1	Do Nothing					
2	DMR Only					
3	P25 Only					
4	TETRA Only					
5	DMR + TETRA					
6	P25 + TETRA					
7	LTE Only					
8	LTE Shared					
9	Commercial LTE					
10	P25 + LTE					
11	TETRA + LTE					
12	P25 + TETRA + LTE					
13	Other Consideration					

constructed where available technology considerations can be listed, and a qualitative scoring (low, medium, and high) can be assessed for each criterion established already.

The alternatives that include more than one technology may be preferred for various reasons. For example, there may be existing deployment based on a technology that may not satisfy the established criteria and requirements. In this case, while deploying a new system based on another technology, the existing system may continue to serve alongside the other technology for the foreseeable future. Another reason for the dual technology consideration could be that one technology may not satisfy all the requirements. Therefore, the two technologies jointly provide full coverage of all the requirements.

However, it is clear that the LTE technology-based approach should be the long-term goal. All immediate, intermediate, and long-term activities should be planned based on this primary objective in mind. A roadmap must be developed toward the realization of this objective.

1.5.3 Economic and Financial Considerations

The cost of planning, deploying, and operating a critical communication system is very high. This is especially much higher for nationwide critical communications networks; the estimated costs of broadband-based public safety networks around the world are running over many billions of dollars. Naturally, substantial costs are surrounded by substantial cost uncertainties. Therefore, accurate cost estimations and fine points of financing are critically important. Specific mathematical models can be beneficial in minimizing potential errors and should be used before the implementation stage [70–75].

Government projects are not usually profit driven and are undertaken for the good of society, and the costs are covered by the government budget [62]. Therefore, many government projects are financed by the taxpayers and are not subject to the classic cost-benefit analysis performed by commercial corporations. However, it is strongly suggested that governments implement general corporate financing rules as much as possible in evaluating their projects to ensure that optimality is achieved in initiating and managing projects.

Financing a critical communications network is challenging; government-operated public safety networks are especially a complicated issue mostly due to the size of financing and integrated ongoing operating cost of the project. No government can easily include a line item in billions of dollars without proper planning and preparation and without disturbing the ongoing operations of the government. A project of this magnitude would require the use of all possible forms of financing including government bonds, equipment leasing, vendor financing, private partnerships, and sharing the network with utility companies and the like [74, 76]. Specific taxes on harmful line items may also be considered; among them are cigarettes and alcohol.

This book provides a detailed discussion of the cost-benefit structure. An analysis by a European economic center indicates that the socioeconomic benefits computed for European Union countries would be approximately 34 billion Euros, annually. In contrast, the opportunity cost of the above scenario for the European Union is to sell the spectrum at an auction to obtain a one-off economic gain totaling 3.7 billion Euros [2, 77]. Naturally, the benefits are several times greater than the opportunity cost, suggesting there that there should be no doubt in implementing broadband public safety networks. The government must educate all involved parties about the socioeconomic benefits of broadband public safety networks.

A proper cost-of-capital estimation also helps the government in the planning and financing stages as it creates a reference value, a benchmark to compare alternatives. With an accurately computed cost of capital value, the government will negotiate better.

1.5.4 Paving the Way

However, before we design, develop, and operate a critical communications system, we need to have a high level, but clear understanding of what needs to be done. We are referring to policy and institutional framework, which should include the following:

- An overall *communications policy* for the entire organization or the whole country, whichever applies; the plan should address commercial, public, government, and military needs and interests. The critical communications system must be an integral part of this plan. The National Broadband Plan prepared by the FCC is an example of this case [78]. If this plan does not exist, it must be developed before launching a critical communications project.
- A separate *authority* with full and overall responsibility to build and maintain the critical communications system, to handle the coordination among all the stakeholders and users, and to make the necessary adjustments and improvements to the network as conditions change and technology evolves.
- In the case of a public safety network, a comprehensive evaluation of potential spectrum alternatives to support a new public safety communications system must be performed. Sufficient bands (at least $2 \times 10\,\text{MHz}$) in $700\,\text{MHz}$ and $800\,\text{MHz}$ must be considered for public safety broadband spectrum as well. Unique attributes of each of these bands, including some technical and regulatory issues, need to be carefully considered in this evaluation.

Building a new critical communication system that is flexible and adaptable to changing needs is a significant challenge. The book recommends a gradual (as opposed to a top-down) approach, which requires that the system be built incrementally and iteratively; each increment is tested under the most possible realistic conditions by the stakeholders involved before the next increment is handled. Next, the book recommends the establishment of a framework for a *national test bed* in which implementations of a new system or subsystem can be validated before they are put into service.

1.5.5 Design and Deployment

Before the deployment of the critical communications system, a high-level *network architecture*, followed by a detailed *network design* must be prepared. As part of this effort, an outreach program must also be developed and executed to gain the maximum level of acceptance from all stakeholders. The processes for supply acquisition must be determined and carried out.

Network Architecture is a framework for the specification of the components and the configuration of a communications system. It is a blueprint that is utilized in developing a detailed network design and in deploying the network. Usually, it is composed of two primary documents: a functional architecture consisting of the functions necessary and needed for the network and a physical architecture, where the functional entities are mapped into corresponding physical counterparts. The operational principles and procedures, as well as the data formats used in its operation, may be a part of these architecture specifications.

Luckily, general network architectures for the well-known critical communications technologies are well-specified and documented. Chapter 12 provides a more detailed discussion on these.

In the context of a public safety network deployment, LTE can follow some structural variations, mainly because there are many existing commercial LTE deployments in almost every country.

- **Private Public Safety LTE Network**—a private LTE RAN and core network infrastructure for the sole purpose of public safety services.
- **Hosted Core Public Safety LTE Network**—public safety entities share a common core network that services their own private LTE eNodeBs.
- **Shared Commercial Public Safety LTE Network**—public safety agencies use commercial LTE networks for public safety services.

Each of these options has advantages and disadvantages, technically, financially, as well as politically [79]. However, a typical implementation of an LTE network is composed of five distinct segments: e-UTRAN, transport, EPC, applications, and operations support systems. Each of these segments presents a different set of challenges in design and deployment.

In the *e-UTRAN segment*, the determination of the number of eNBs (LTE base stations) and their locations are critically important in the design of the network. The necessary information impacting coverage and capacity design must be identified, collected, processed, and calculated. The book provides extensive details about the type of information needed in this context [80].

In the *transport segment*, a backhaul network must be designed to carry traffic from eNBs to the elements in the core network. The high-level design should focus on the transport media, transport technology, and the topology. To meet LTE requirements, the fiber as a transport media, Layer 2 as a transport technology, and the ring topology is highly recommended [81].

In designing the EPC, the *core network*, a critical decision is to determine whether to deploy a centralized or distributed architecture. A distributed model is favorable because of the importance of network availability. The LTE core network design also includes core network dimensioning, which is used to determine the number of nodes and the capacity required.

Applications to be used by the users and agencies must be a part of the design and planning of the LTE network and its sites. Some applications require high bandwidth capacity while others involve real-time transmission. Typical applications that can be used in the public safety sector are video, dispatch, fingerprint, image transfer, voice over IP, push to talk, mobile database query, machine to machine, and monitoring.

Operation Support Systems (OSSs), which are used to keep the network up and running, and providing its services satisfactorily as "promised," must be a part of the overall design and planning of the public safety network. OSS functionalities include fault management, configuration management, accounting management, performance management, and security management (FCAPS).

After the architectural concepts discussed above are approved, a detailed **network design**, also called *low-level network design*, must be prepared. A detailed design document describes how the network infrastructure should be built and engineered to meet the specific goals and objectives delineated in various documents, including the network architecture. Usually, the detailed design includes every single bit of information that is necessary for building and deploying the network. For example, the identification of the switch ports that need to be connected to the router should be specified in a detailed network design.

Designing a critical communications system requires engineers knowledgeable in cellular network design since there are certain similarities such as planning for the cellular sites and deployment, a radio network, back-haul transmission, and the core network. However, there are several significant differences especially in coverage and capacity (radio network dimensioning is mostly coverage driven by a critical communications network). For example, a critical communications system uses group calls with only one channel per group, but several sites in one call. The average call duration is much shorter. Unlike a commercial network, the additional traffic that dispatch stations and command systems generate and receive needs to be incorporated in critical communications systems.

Once the planning and design phase is completed and approved, activities in deploying the network should begin. Like the design phase, the deployment phase also includes careful considerations for each segment of the overall network. Before the actual deployment begins, a *deployment plan* must be developed to include installation, integration, and test procedures for the nodes to be deployed. It is most likely that the deployed systems will include equipment from a number of different vendors. Therefore, it is crucial to perform an end-to-end system integration testing to verify the requirements established during the design and procurement phase. When the system integration is completed, a verification testing must be performed to verify stability, media quality, robustness, maintainability, capacity, and coverage. All these are explained in greater detail in the book.

1.5.6 Operations, Administration, Maintenance, and Provisioning (a.k.a. Management)

Once the network is deployed, and integration and verification tests are performed, the network is ready to provide services to its users. For the system to work correctly, there needs to be a "network and service management infrastructure" in place. This includes a set of OSSs, applications, plans, policies, procedures, and people. This area is crucial for a critical communications system since in extreme situations—on-scene operations—the system must be extremely resilient and must be up to help the first responders. An *operations plan* to describe the resources, organizations, responsibilities, policies, and operations procedures to monitor and manage the network efficiently must be developed. The operations plan and procedures are executed by staff members of a Network Operations and Control Center (NOCC), which is set up to monitor, control, and manage all segments of the network. In either case, it is crucial to implement the same type of operation procedures. It is highly recommended that the NOCC organization and operations models be aligned with the standard enhanced Telecommunications Operations Map (eTOM) defined by the Telecommunications Management Forum (TMF) and the Information Technology Infrastructure Library (ITIL) to synchronize the activities among geographically dispersed regional centers when applicable [82, 83].

The term *Operations Support System (OSS)* is a generic term used to refer to the systems used in operating, administering, maintaining, and provisioning the networks. Depending on the size of the network, there could be many OSSs—regional, national, specialized (e.g. billing), general purpose, etc. Moreover, there may be more specific names used to signify the specific purpose that an OSS is used for. The Telecommunication Management Network (TMN) provides a framework to name OSSs more formally. Briefly, TMN defines management layers (business, service, network, and element) and names OSSs according to the layer in which they are used. For example, the OSS used at the service management layer is called the Service Management System (SMS). Accordingly, the OSS used at the network management layer is called Network Management Systems (NMS)[84] (the term NMS is also used generically especially in smaller, data specific networks [e.g. LANs] to refer to the management workstations used in managing [i.e. operating, administering, and maintaining] networks).

In critical communication networks too, OSSs are used to enable configuration, management, and maintenance of all network elements (e.g. switches, base stations, dispatcher consoles, and links). Both Project 25 and TETRA networks have standardized interfaces to OSSs, which typically use the Simple Network Management Protocol (SNMP) to collect network management information and alarms (note that both Project 25 and TETRA specifications use the term NMS, rather than OSS). Depending on the size of the network, there could be a hierarchy of OSSs. For example, a low-level OSS (e.g. SNMP Console Manager) may report the information collected to higher-level systems for further processing and displaying. Note that an SNMP Console Manager can request information and the status of one or more alarms from any network element.

The OSS typically records all network events in a database or files. Date, time, and source, together with the event type, are recorded to enable system reporting including network traffic loading and usage and timeslot distribution. Daily files can be further processed by specialized systems or manually in spreadsheets to provide detailed statistical analysis for performance management reporting and system optimization.

1.6 SUMMARY AND CONCLUSIONS

This book deals with the technologies, systems, and applications used in public safety and mission-critical communications specific operations. The book covers economic, financial, and policy issues as well as the design, deployment, and operation of such systems.

Critical communications networks provide the basis for *situational awareness* and *command and control* capabilities, which roughly translate into the delivery of mission-critical data, survivability against multiple failures, maintenance of data integrity and confidentiality, essential full coverage and capacity, interoperability with other networks, and required support for officers, applications, and devices.

As of writing this book, most old analog technologies used for critical communications systems have been replaced by *all* digital narrowband technologies led by Project 25, TETRA, and DMR standards. There is also a higher consensus that LTE be the technology of the future for critical communications systems. TETRA has been the choice of public safety agencies mainly in Europe and Project 25 technologies mainly in North America, but both have worldwide deployments as well. DMR-based systems have also been chosen in some regions, although not as extensively as TETRA and Project 25.

Project 25 and TETRA technologies are mature, widely used, tested, reliable, and feature rich in voice applications. These narrowband technologies are somewhat limited in providing data services. Also, narrowband technologies are more expensive since the target market is limited, compared to the commercial mobile market. The demand for data-intensive applications by public safety agencies is increasing.

LTE technology is an ideal candidate for a nationwide critical communications system, especially for public safety applications. It is a proven and tested technology for commercial use and nationwide broadband networks. It is handling broadband data applications an order of magnitude better than the narrowband systems. For the first time in history, LTE has emerged as a single worldwide standard and is used commercially everywhere around the world. The scale of economy is just outstanding. A growing number of countries, including the USA, have chosen LTE for their public safety networks already.

A complete critical communication system encompasses applications supported by a set of comprehensive systems, purpose-built, intuitive devices, and comprehensive services. While applications are commonly deployed over the Internet, application developers have traditionally been unable to produce packet data applications over TETRA and Project 25 due to the low data rate provided. However, there have been some offerings by various vendors to ease the concern somewhat. There are some applications currently available for various markets including, but not limited to, police, fire, ambulance, transport, airport, field service, and utilities.

To serve all the users in a coordinated way, critical communications systems usually have some centers and associated support systems. Two of the most important ones are incident management systems, which enable all the users and agencies to work together to handle incidents that are reported or detected, and operations and control systems, which are used to operate, administer, and maintain the network.

Terminal devices for the users of critical communications systems strictly depend on the underlying communications being used. For example, the user devices for TETRA technology will be different from the user devices for Project 25 technology. Similarly, LTE-based critical communications devices will be drastically different from its narrowband counterparts, handling and displaying multimedia, just like the smartphones and tablets used commercially. Regardless of the technology being used, end-user devices can be roughly categorized as mobile radios, portable radios, and consoles.

As expected, spectrum issues are being addressed in many countries around the world.

Each region has a slightly different approach. For example, in the USA, most voice land mobile radio systems use narrowband frequencies in the VHF and UHF bands. However, the FCC has recently allocated 758–768 MHz and 788–798 MHz for base stations and mobile units use, respectively.

Standardization of critical communications interfaces and protocols has been handled mainly by the TIA for Project 25 and by the ETSI for TETRA and DMR-related projects. The standardization work on LTE-based standards has been carried out mainly by 3GPP, a collaboration among groups of telecommunications standards associations.

Planning, designing, and deployment of a critical communications system depend on many factors including whether it is for a country or a commercial company, whether it will be a nationwide or a regional system, and whether there is already an existing system in place. Although the level of effort, activities, and contents heavily depends on the factors mentioned above, there are some common activities in planning for a critical communications system. A feasibility study needs to be conducted in developing a formal business plan. A typical feasibility study should include operational feasibility, market feasibility, financial/economic feasibility, organizational/managerial feasibility, environmental feasibility, and legal feasibility.

An essential part of the planning effort is to select a technology for the planned critical communications systems. Currently, there are four major candidate technologies to consider: TETRA, Project 25, DMR, and LTE. Alternative scenarios that include more than one technology may be preferred for various reasons such as to fulfill the requirements outlined in the planning step. A consensus is that the LTE technology-based approach is the long-term goal.

The cost of planning, deploying, and operating a critical communications system is very high, running over several billion dollars. Naturally, substantial costs are surrounded by substantial cost uncertainties. Therefore, accurate cost estimations and fine points of financing are critically important. Specific mathematical models can be beneficial in minimizing potential errors and should be used before the implementation stage. A project of this magnitude would require the use of all possible forms of financing including bonds, equipment leasing, vendor financing, private partnerships, sharing the network with utility companies, and the like.

Before the deployment of the system, a *network architecture*, followed by a detailed *network design* must be prepared. The processes for supply acquisition must be determined and carried out. Once the planning and design phase is completed and approved, activities in deploying the network should begin. Before the actual deployment begins, a *deployment plan* must be developed to include installation, integration, and test procedures for the nodes to be deployed. It is crucial to perform an end-to-end system integration testing to verify the requirements established during the design and procurement phase.

For the system to work correctly, there needs to be a "network and service management infrastructure" in place. An *operations plan* to describe the resources, organizations, responsibilities, policies, and operations procedures to monitor and manage the network efficiently must be developed. This area is crucial for a critical communications system since in extreme situations, that is on the scene operations, the system must be extremely resilient and must be up to help the first responders.

REFERENCES

1. TETRA and Critical Communications Association, "Broadband spectrum for mission critical communication needed," Position Paper, Aug. 2013.
2. B. Mattsson, "TETRA News—TETRA evolution for future needs," *TETRA Applications*, Mar. 17, 2014. [Online]. Available: http://www.tetra-applications.com/27885/news/tetra-evolution-for-future-needs. [Accessed: Jul. 24, 2016].
3. National Public Safety Telecommunications Council, "Defining public safety grade systems and facilities final report," A NPSTC Public Safety Communications Report, May 2014.

4. A. Bleicher, "LTE-Advanced is the real 4G," *IEEE Spectr.*, Dec. 31, 2013. [Online]. Available: https://spectrum.ieee.org/telecom/standards/lte-advanced-is-the-real-4g.

5. C. Cox, *An Introduction to LTE: LTE, LTE-Advanced, SAE, VoLTE and 4G Mobile Communications*, 2nd ed. John Wiley & Sons, 2014.

6. S. Burfoot, P. Chan, and D. Reitsma, "P25 radio systems training guide, revision 4.0.0," Codan Radio Communications, 2013.

7. J. Oblak, "Project 25 phase II," EFJohnson Technologies, Dec. 2011.

8. The Network Encyclopedia, "Terrestrial Trunked Radio (Tetra) in The Network Encyclopedia," 2013. [Online]. Available: http://www.thenetworkencyclopedia.com/entry/terrestrial-trunked-radio-tetra/. [Accessed: Jul. 19, 2016].

9. P. Stavroulakis, *TErrestrial Trunked RAdio—TETRA*. Springer Science & Business Media, 2007.

10. J. Dunlop, D. Girma, and J. Irvine, *Digital Mobile Communications and the TETRA System*. John Wiley & Sons, 2013.

11. Tetrapol, "Tetrapol forum." [Online]. Available: http://www.tetrapol.com/. [Accessed: Jul. 24, 2016].

12. "DMR versus TETRA systems comparison Version 1v2," *Radio Activity Solutions*, 2009.

13. B. Bouwers, "Comparison of TETRA and DMR," Rohill Technologies, 2010.

14. R. Marengon, "A TETRA and DMR comparison," *Radio Resource Magazine*, Apr. 1, 2010. [Online]. Available: https://www.rrmediagroup.com/Features/FeaturesDetails/FID/174.

15. "The DMR standard," *Digital Mobile Radio Association*, Dec. 29, 2010. [Online]. Available: http://dmrassociation.org/the-dmr-standard/. [Accessed: Jul. 24, 2016].

16. Signals and Systems Telecom, "The public safety LTE & mobile broadband market: 2012–2016," Market Report, Nov. 2012.

17. "TETRA and LTE working together v1.1." TETRA and Critical Communications Association, Jun. 2014.

18. Project MESA, Service Specification Group—Services and Applications, "Statement of requirements executive summary." [Online]. Available: http://www.projectmesa.org/MESA_SoR/mesa_sor_executive_summary.pdf. [Accessed: Sep. 13, 2018].

19. WiMAX Forum, "AeroMACS, WiGRID, and WiMAX advanced technologies." [Online]. Available: http://www.wimaxforum.org. [Accessed: Jul. 24, 2016].

20. Motorola Solutions, "The future is now: Public safety LTE communications," Motorola Solutions White Paper, Aug. 2012.

21. Tait Limited, "Introducing unified critical communications," White Paper, 2014. [Online]. Available: https://go.taitradio.com/introducing-unified-critical-communications-for-public-safety.html. [Accessed: Sep. 13, 2018].

22. B. Deverall, Added Value Applications, "Briefing for E-Gif Working Group on incorporating the APCO P25 standards into the e-government interoperability framework," Sep. 19, 2006.

23. "Comments of internet 2 to NTIA," Counsel for Internet 2, Docket No. 120928505-2505-01, Nov. 2012.

24. The Critical Communications Association (TCCA), "TCCA signs market representation partner agreement with 3GPP." [Online]. Available: https://tcca.info/tcca-signs-market-representation-partner-agreement-with-3gpp/. [Accessed: Sep. 13, 2018].

25. TETRA and Critical Communications Association, "Mission Critical Mobile Broadband: Practical standardization & roadmap considerations," White Paper, Feb. 2013.
26. D. Jackson, "UK seeks to replace TETRA with LTE as early as 2016," *Urgent Communications*, Jun. 6, 2013. Available: http://urgentcomm.com/tetra/uk-seeks-replace-tetra-lte-early-2016.
27. E. Dahlman, S. Parkvall, and J. Skold, *4G LTE/LTE-Advanced for Mobile Broadband*. Academic Press of Elsevier, 2011.
28. National Telecommunications Information Agency, "Desirable properties of a national public safety network," Report and Recommendations of the Visiting Committee on Advanced Technology, Jan. 2012.
29. I. Sharp, "Delivering public safety communications with LTE," 3rd Generation Partnership Project, Jul. 2013.
30. TCCA, "TCCA liaison to 3GPP SA on group communications and proximity services," 3GPP SP-120456, Jul. 2012.
31. 3rd Generation Partnership Project, "3GPP release 12 LTE," Mar. 2015. [Online]. Available: http://www.3gpp.org/specifications/releases/68-release-12. [Accessed: Sep. 13, 2018].
32. Alcatel-Lucent, "Alcatel-Lucent and first responders conduct trial of 4G LTE public safety broadband mobile network," Alcatel-Lucent Press release, Nov. 2013.
33. National Telecommunications and Information Administration, "FirstNet." [Online]. Available: http://www.ntia.doc.gov/category/firstnet. [Accessed: Sep. 13, 2018].
34. C. Gessner, *Long Term Evolution: A Concise Introduction to LTE and its Measurement Requirements*. Rohde & Schwarz Publication, 2011.
35. W. Lehr and N. Jesuale, "Spectrum pooling for next generation public safety radio systems," in *IEEE Proc. Dynamic Spectr. Access Netw. (DySPAN2008) Conf.*, Chicago, Oct. 2008, pp. 1–23.
36. 5G Americas, "List of 3G/4G deployments worldwide (HSPA, HSPA+, LTE." [Online]. Available: http://www.4gamericas.org/en/. [Accessed: Jul. 24, 2016].
37. A. M. Seybold, "Seybold's take: Public safety's 700 MHz LTE network an opportunity for vendors," *Fierce Wireless*, Mar. 14, 2012. Available: https://www.fiercewireless.com/wireless/seybold-s-take-public-safety-s-700-mhz-lte-network-opportunity-for-vendors.
38. M. Poikselkä, Harri Holma, Jukka Hongisto, Juha Kallio, and Antti Toskala, *Voice over LTE (VoLTE)*. John Wiley & Sons, 2012.
39. Etherstack, "PMR-LTE network solutions: Push-to-talk PMR over LTE," White Paper, 2013.
40. K. Doppler, M. Rinne, C. Wijting, C. Ribeiro, and K. Hugl, "Device-to-device communication as an underlay to LTE-Advanced networks," *IEEE Commun. Mag.*, vol. 47, no. 12, pp. 42–49, Dec. 2009.
41. P. Janis, C.-H. Yu, K. Doppler, C. Ribeiro, C. Wijting, K. Hugl, O. Tirkkonen, and V. Koivunen, "Device-to-device communication underlying cellular communications systems," *Int. J. Commun. Netw. Syst. Sci.*, vol. 2, no. 3, pp. 169–178, 2009.
42. Qualcomm, "Study on LTE Device to device proximity discovery," 3rd Generation Partnership Project, TSG RAN Meeting #57, 2012.
43. 3rd Generation Partnership Project, "Mobile broadband standard: LTE." [Online]. Available: http://www.3gpp.org/technologies/keywords-acronyms/98-lte. [Accessed: Jul. 24, 2016].

44. *Wikipedia, the free encyclopedia*, "3GPP standards," Jun. 30, 2016. Available: https://en.wikipedia.org/wiki/3GPP.

45. J. Liu, Y. Kawamoto, H. Nishiyama, N. Kato, and N. Kadowaki, "Device-to-device communications achieve efficient load balancing in LTE-Advanced networks," *IEEE Wirel. Commun.*, vol. 21, no. 2, pp. 57–65, Apr. 2014.

46. 5G-PPP, "5G-PPP." [Online]. Available: https://5g-ppp.eu/. [Accessed: Jul. 24, 2016].

47. US Department of Homeland Security SAFECOM, "Public safety statement of requirements for communications & interoperability," vol. 1, Version 1.2, Oct. 2006.

48. P. R. Kempkerhttp, Department of Homeland Security, "Basic gateway overview." [Online]. Available: http://www.c-at.com/Customer_files/Fairfax%20rally/Module%201%20-%20Basic%20Gateway%20Overview.ppt. [Accessed: Sep. 13, 2018].

49. E. Olbrich, "Public safety communications research public safety LTE," presented at the LTE World Summit, Barcelona, May 2012.

50. D. Witkowski, "Effectively testing 700 MHz public safety LTE broadband and P25 narrowband networks," Anritsu Company, 2013.

51. D. Tuite, "Can public-safety radio's P25 survive LTE?." [Online]. Jul. 17, 2012. Available: https://www.electronicdesign.com/analog/can-public-safety-radio-s-p25-survive-lte.

52. EU CORDIS, "GERYON (Next generation technology independent interoperability of emergency services)," 284863.

53. APCO International, "APCO Application Community." [Online]. Available: http://appcomm.org/. [Accessed: Jul. 25, 2016].

54. Team Simoco, "TETRA-GTI applications." [Online]. Available: http://www.simocogroup.com/resources/product_datasheets/tetra/tetra-gti/TETRA-GTI_Applications.pdf. [Accessed: Jun. 25, 2016].

55. A. M. Seybold, "FirstNet brings changes to Unified Incident Command," *IMSA J.*, Feb. 2014.

56. J. W. Morentz, C. Doyle, L. Skelly, and N. Adam, "Unified Incident Command and Decision Support (UICDS) a Department of Homeland Security initiative in information sharing," in *Proc. IEEE Conf. Technol. Homeland Security (HST '09)*, Boston, May 2009, pp. 182–187.

57. Motorola Solutions web site. [Online]. Available: http://www.motorolasolutions.com/en_us/products/two-way-radios.html. [Accessed: Jul. 25, 2016].

58. Public Safety Spectrum Trust. [Online]. Available: http://www.psst.org/. [Accessed: Jul. 18, 2016].

59. Carlos M. Gutierrez, Meredith A. Baker, US Department of Commerce, National Telecommunications Information Agency, "Spectrum management for the 21st century: The President's spectrum policy initiative—Federal strategic spectrum plan," Mar. 2008. [Online]. Available: http://www.ntia.doc.gov/reports/2008/FederalStrategicSpectrumPlan2008.pdf. [Accessed: Sep. 13, 2018].

60. Federal Communications Commission, "FCC takes action to advance nation-wide broadband communications for America's first responders," Jan. 2011. [Online]. Available: http://hraunfoss.fcc.gov/edocs public/attachmatch/DOC-304244A1.pdf. [Accessed: Sep. 13, 2018].

61. "Federal Communications Commission, 700 MHz band plan for public safety services." [Online]. Available: https://www.fcc.gov/general/700-mhz-public-safety-spectrum-0. [Accessed: Sep. 13, 2018].

62. TETRA and Critical Communications Association, "Spectrum saves lives: High speed data for the police and emergency services," presented at the The Operators Workshop, 2011.

63. "Project 25: What's next for the Global Standard?," *Mission Critical Communications Magazine*, Oct. 2013.

64. "TETRA + Critical Communications Association." [Online]. Available: http://www .tandcca.com/. [Accessed: Jul. 24, 2016].

65. "3GPP." [Online]. Available: http://www.3gpp.org/. [Accessed: Jul. 24, 2016].

66. R. Favraud, A. Apostolaras, N. Nikaein, and T. Korakis, "Towards Moving Public Safety Networks," *IEEE Commun. Mag.*, vol. 54, no. 3, pp. 14–20, Apr. 2016.

67. H. Hoagland and L. Williamson, *Feasibility Studies*. [Online]. University of Kentucky, 2000. Available: http://www.parcodelfuenti.it/dossier/documenti/Dispensa.Univ.Kentucky .pdf.

68. H. Tompson, "Business feasibility studies: Dimensions of business viability," Best Entrepreneur, 2003.

69. W. Truitt, *A Comprehensive Framework and Process*. London: Quorum Books, 2003.

70. R. Hallahan and J. M. Peha, "Quantifying the costs of a nationwide broadband public safety wireless network," in *Proc. 36th Telecommun. Policy Res. Conf.*, Arlington, Sep. 2008.

71. R. Hallahan and J. M. Peha, "The business case of a nationwide wireless network that serves both public safety and commercial subscribers," in *Proc. 37th Telecommun. Policy Res. Conf.*, Arlington, Sep. 2009.

72. R. Hallahan and M. Peha, "Compensating commercial carriers for public safety use: Pricing options and the financial benefits and risks," in *Proc. 39th Telecommun. Policy Res. Conf.*, Arlington, Sep. 2011.

73. Federal Communications Commission, "The public safety nationwide interoperable broadband network: A new model for capacity, performance and cost," White Paper, Jun. 2010.

74. J. M. Peha, "A public-private approach to public safety communications," *Issues Sci. Technol.*, vol. 29, no. 4, 2013.

75. N. Bolari, "Indirect returns and use of NPV in financial viability modelling of critical communications networks," *IEEE Commun. Mag.*, vol. 54, no. 3, pp. 38–43, Mar. 2016.

76. Utilities Telecom Council, "Sharing 700 MHz public safety broadband spectrum with utilities: A proposal." [Online]. Oct. 2012. Available: https://utc.org/wp-content/uploads/ 2018/02/Sharing-700-MHz-Public-Safety-Broadband-Spectrum-With-Utilities-2.pdf.

77. A. Grous, "Socioeconomic value of mission critical mobile applications for public safety in the EU: 2 × 10 MHz in 700 MHz in 10 European countries," Centre for Economic Performance, London School of Economics and Political Science, Dec. 2013. Available: http://eprints.lse.ac.uk/69180/1/Grous_Socioeconomic_value_of_mission_critical_ applications_UK_2013_author.pdf.

78. Federal Communications Commission, "Connecting America: The national broadband plan." [Online]. Mar. 2010. Available: https://transition.fcc.gov/national-broadband-plan/ national-broadband-plan.pdf.

79. Federal Communications Commission, "Public safety tech topic #22—Application of emerging wireless broadband technology for public safety communications." [Online].

Available: https://www.fcc.gov/help/public-safety-tech-topic-22-application-emerging-wireless-broadband-technology-public-safety. [Accessed: Sep. 13, 2018].

80. S. Palat and P. Godin, "LTE network architecture: A comprehensive tutorial," Alcatel Lucent, Strategic White Paper, 2009.

81. Aviat Networks, "Five recommendations for building FirstNet-ready backhaul networks," White Paper, Aug. 2013.

82. TM Forum, "Business Process Framework (eTOM)." [Online]. Avaialble: https://www.tmforum.org/business-process-framework. [Accessed: Sep. 13, 2018].

83. "IT Infrastructure Library (ITIL)." [Online]. Available: https://www.itgovernanceusa.com/itil. [Accessed: Jul. 25, 2016].

84. S. Aidarous and T. Plevyak, eds., *Telecommunications Network Management into the 21st Century*. IEEE Press, 1994.

2

USERS OF CRITICAL
COMMUNICATIONS SYSTEMS

This chapter discusses the users of critical communications systems. The types of users, what they do, and what types of communications systems they use currently are discussed.

2.1 INTRODUCTION

To procure, design, and deploy critical communications systems so that they operate satisfactorily, one must know its users, the organizations where the users are employed, and the tasks the users perform to do their jobs.

We can group the users of critical communications systems into roughly two categories:

- The employees of the public (government and volunteer) agencies involved in public safety—law enforcement agencies and first responders such as fire departments and ambulatory services.

Fundamentals of Public Safety Networks and Critical Communications Systems: Technologies, Deployment, and Management,
First Edition. Mehmet Ulema.
© 2019 by The Institute of Electrical and Electronics Engineers, Inc. Published 2019 by John Wiley & Sons, Inc.

- The employees of commercial and public organizations where communications systems are mission critical. Other organizations needing communications systems include those involved in transport, utilities, mining, and similar businesses.

The following sections discuss the organizations where users are employed to perform public safety-related activities. The focus of the discussions below is on organizations, the functions provided, the activities performed, the types of vehicles driven, and the communications systems used.

2.2 ORGANIZATIONS INVOLVED IN PUBLIC SAFETY

These are the agencies involved in public safety affairs. Typical functions include first and emergency responses to wide-scale natural disasters such as earthquakes, forest fires, and flooding, and man-made disasters such as nuclear explosions, radiation, terrorism, as well as localized emergencies such as automobile accidents, fires, medical emergencies, and any other threats to public order. A police department is a typical example of a public safety agency.

The term "first responders" is an umbrella term that refers to those agencies that are designated and trained to respond to emergencies. They include fire, police, and emergency medical personnel. The Coast Guard, National Guard, and gendarmeries can also be included in this category since they are also trained to respond to emergencies, especially those on a larger scale, in some areas. It is essential that they have access to reliable, interoperable communications to assist those in need during emergencies in a coordinated way.

The term "law enforcement" is another umbrella term that refers to those agencies that are involved in enforcing laws. The term typically covers the police, Coast Guards, gendarmeries, and other similar agencies.

Disaster relief agencies, like the Federal Emergency Management Agency (FEMA) in the USA and other emergency management agencies also participate in pre-, during, and post-emergency events, in cooperation with other first responders. Their focus may be a little different. For example, FEMA is involved in providing financial support to those affected by disastrous events.

Last, but not the least, there are nongovernmental organizations (NGOs) like the Red Cross, Red Crescent, Catholic Relief Services, and other similar organizations that are heavily involved in supporting disaster relief efforts worldwide.

Each of these agencies, as well as some other agencies involved in public safety, are discussed in more detail in the following sections.

2.2.1 Police Departments

Police departments are mainly responsible for law enforcement and protection of citizens. Traditional police functions include monitoring, investigating, and responding to criminal activities as well as making arrests to deter illegal and dangerous behavior. Police functions also include responding to emergency situations such as accidents and disasters, controlling traffic, and issuing tickets to dangerous and careless drivers.

Police officers and detectives in police departments are also involved in helping other law enforcement agencies such as the Department of Justice to investigate and prosecute criminals. Their duties also include taking statements from suspects and witnesses as well as collecting evidence from crime scenes to help forensic scientists involved in investigations.

Police departments interact and collaborate with some different organizations such as health care providers, fire departments, the Coast Guard and National Guard, local businesses, the media, religious leaders, and organizations working with victims and communities.

How the police force is organized varies considerably from country to country. In some countries, there is a central organization that controls all police departments around the country. In some countries, police departments are under the township government, e.g. the mayor, and entirely independent of each other, with total responsibility under their jurisdiction. Of course, for those crimes that cross over their jurisdiction, they interact and collaborate with the police force in neighboring towns, the state police, and federal law enforcement agencies as well.

The term "police" is not the only word used to refer to these organizations. Other names used in different countries include constabulary, gendarmerie, police force, security units, police service, crime prevention, protective services, law enforcement agency, civil guard, or civic guard. In addition to "police officer," other terms such as troopers, road agents, sheriffs, constables, rangers, peace officers, or civic/civil guards may also be used [1].

Police departments carry out their functions by using, in addition to traveling by foot and with horses, a variety of vehicles including bicycles, motorcycles, scooters, cars, sport-utility vehicles (SUVs), minivans, buses, boats, helicopters, small airplanes, and armored trucks.

2.2.2 Fire Departments

Fire departments provide firefighting and rescue functions in emergency situations. Firefighting includes structural firefighting and wildland firefighting. Other areas fire departments are involved in include life-saving through search and rescue, and management of hazardous materials.

They are also trained, in many countries, as paramedics, emergency medical technicians, ambulance technicians, etc., so that they can provide some services in medical emergencies such as stabilizing the victim until an ambulance can arrive. They are also involved in fire prevention programs aimed at preventing fire-related accidents to save lives and property.

Depending on the region or country, fire departments could be public or private institutions that may employ volunteers and paid employees trained as firefighters.

Fire departments may also be called the fire service, fire, rescue service, fire brigade, or some local names.

Fire departments typically operate in a specific geographic area (e.g. a village, a town, a city, or a county). There are usually more than one so-called "fire stations" in a given fire department. These stations are primarily used to house firefighting vehicles and offices for the personnel.

The most common vehicles that fire departments use are firefighting engines, also called pumps in some countries. They carry a hose, a pump, and a water tank. Engines can be customized to serve as structural firefighting in urban areas with many high risers or in rural areas.

Fire departments may also use trucks to support firefighting efforts such as search and rescue, and forcible entry. Large aerial ladders and platforms may also be available for use in some fire departments to support firefighting in structural fire. There are some vehicles, called quints, which combine all these, i.e. engine, pump, hose, tank, ladder, and platform.

Fire departments may use several other types of vehicles such as helicopters, firefighting aircrafts, jet boats, ambulances, and even bulldozers.

2.2.3 Emergency Medical Services

Emergency medical services are also known as ambulance services, paramedic services, a first aid squad, emergency squad, ambulance squad, ambulance service, ambulance corps, or life squad. Emergency medical services personnel provide critical care. They handle the first possible medical treatment to the injured. Then they arrange immediate transportation of the injured to the nearest health care facility where complete treatment can be provided in a safe and controlled environment. Emergency medical services may also get involved in moving patients from one facility/hospital to another one, which may be better equipped to handle some individual cases.

Like fire departments, emergency medical services may be organized in a variety of ways depending on the country, state, province, county, cities, and townships. Some could just be a voluntary organization supported by volunteers in that area. In some cases, they can operate as a division of the police or fire departments.

The personnel providing emergency medical services are typically cross-trained to aid the injured as a first treatment as well as to participate in rescue operations. Drivers, low-level to high-level technicians, nurses, and even medical doctors make up the personnel employed in providing emergency medical services.

Ambulances are the most common vehicle type used by emergency medical service departments. They may vary by size (storage and payload capacity), communication capabilities, and the medical facilities they contain.

Like other first responders, emergency medical services need to be supported by an efficient and rapid dispatching system including fast location finding and accurate navigation capabilities. Furthermore, during the treatment of the injured, the personnel may need to communicate with nearby healthcare facilities and specialist doctors. These communications may require transmission of various diagnosis data including high-resolution images.

2.2.4 Emergency Management Agencies

Emergency management is primarily involved in developing and executing plans to help reduce the impact of disasters, which may include mortality as well as financial loss to individuals, families, and communities. Emergency management agencies are called disaster management, or disaster recovery in some countries.

The departments or agencies involved in devising such plans and getting involved during and after such disasters are called different names depending on the region and country. They are more focused on providing financial support for rebuilding and relief to those individuals and infrastructure that were affected by the disastrous events. Activities of the emergency management agency are handled in cooperation with other first responders.

There may be included local agencies as well as national emergency management agencies. For example, in the USA, FEMA [2], which is a part of the Department of Homeland Security, coordinates the emergency response activities for large-scale disasters that require massive emergency efforts that may be way beyond local and regional capabilities and resources.

2.2.5 Coast Guard

A coast guard (aka coastal guard) is a maritime security organization. A coast guard, typically a part of a country's military services, is responsible for protecting "ports, inland waterways, along with the coasts, and on international waters" [3]. A typical coast guard, in the areas mentioned above, provides safety and security to the people; protects the marine transportation system and infrastructure; and maintains the territorial integrity. A coast guard's responsibilities vary significantly from country to country, "from being a heavily armed military force with customs and security duties to being a volunteer organization tasked with search and rescue functions and lacking any law enforcement powers" [4]. Some landlocked countries such as Switzerland may not have a coast guard organization. However, similar functions in inland waters (e.g. lakes) are provided by specialized military forces or civilian-only organizations.

In addition to military personnel, a coast guard also employs civilian professional as well. For example, the US Coast Guard employs about 40,000 active and reserve military and about 8,000 civilian personnel [3]. Depending on the size of the coastal and inland waters of the country, a coast guard may cover the slightly larger area. The coast guard may use large and small boats on water, and airplanes and helicopters in the air.

As one may observe, the communications systems employed by different coast guards may vary significantly depending on the responsibilities, size, geographical coverage, etc. For example, the Coast Guard in the USA has eight primary radio stations covering long-range transmissions and an extensive network of VHF radio stations along the nation's coastline and inland rivers. The current critical communications system is called Rescue 21, which is an advanced maritime Computing, Command, Control, and direction-finding Communications (C4) system designed to better locate mariners in distress and save lives and property at sea and on rivers [5].

Coast guards may be involved in responding to natural disasters such as hurricanes and man-made disasters such as offshore drilling-related oil spills. Accordingly, the critical communications systems used by coast guards need to be adaptable and dynamically configurable to become interoperable with nearby public safety systems to handle these types of emergency situations. For example, right after the Gulf Coast oil spill began, the Coast Guard, using its interoperable radio network, was able to communicate with public safety agencies in Mississippi, Alabama, and Texas, basically creating an interoperable regional network based on Project 25 digital 700/800 MHz networks [6].

2.2.6 Other Organizations in Public Safety

In addition to the typical organizations discussed above, there are other governments and volunteer organizations that are involved in public safety and use critical communication systems. The following paragraphs provide brief introductions to a few of such organizations.

2.2.6.1 Red Cross and Red Crescent Societies. These NGOs are a network of independent volunteer organizations worldwide involved in disaster relief activities. Each country has one Red Cross or Red Crescent society, typically based on the dominant religion in the country. Although the disaster relief activities of these societies are subject to the local laws where the disaster takes place, these societies are autonomous and adhere to the principles of openness and humanitarian work only [7].

Although Red Cross and Red Crescent organizations get involved in a variety of humanitarian activities, they are well known for their participation in disaster response and recovery situations. Volunteers play a crucial role in their activities. Since they are local and relatively close to the disaster areas than most other agencies, these volunteers are typically the first to be present in these stressed areas.

As in the case of any disaster response and recovery operation, a reliable and efficient communications system is critically important in the provision of effective disaster services by the Red Cross and Red Crescent organizations as well. The availability of a communications system is not only essential for the volunteers to coordinate their activities, but also essential to establish communications channels with the people affected by the disaster. This includes voice, data, and video transmission utilizing radio, satellite, microwaves, and other transmission and telecommunications technologies [8]. To this effect, the Red Cross and Red Crescent organizations collaborate with commercial telecommunication (wireline and wireless) service providers, apps developers, social media companies, as well as other relevant organizations such as the International Amateur Radio Union (IARU) [9].

2.2.6.2 *National Guard.* As we saw before, this name is also used around the world in different contexts to refer to a military organization in some countries, in some other countries to a civilian organization, yet in some countries like the USA to an organization that has both military and civilian components. Some involve mainly military operations, yet some others, like the one in the USA, get involved in military as well as public safety and first responder activities.

In the case of the USA, each state has its own National Guard composed of a reserve military force: They hold a civilian job full-time while serving part-time as a National Guard member. National Guard units can be called into action by the federal government as well to get involved in nationally declared emergency situations, in public safety-related activities, or purely military activities during a war [10, 11].

2.2.6.3 *Gendarmerie.* A gendarmerie is typically a service branch of the military charged with police duties among civilian populations. Modern civilian police forces may retain the title of "gendarmerie" for historical reasons in some countries [12].

The responsibilities and jurisdictions differ from country to country. For example, the Turkish Gendarmerie is mainly responsible for the maintenance of public order generally in rural areas. The Gendarmerie is primarily a law enforcement force of a military nature [13].

2.2.6.4 *US Army Corps of Engineers.* This is an engineering, design, and construction agency, and an Army command (the US Department of Defense) responsible for "delivering vital public and military engineering services; ... to strengthen our Nation's security, ... and reduce risks from disasters" [14]. Their public works include dams, canals, flood control, beach erosion control, dredging for waterways, and ecosystem restoration.

2.3 OTHER SECTORS USING CRITICAL COMMUNICATIONS SYSTEMS

A mission-critical communication system is a network used by organizations to provide communications infrastructure to carry out mission-critical functions. The communications system used by the workers at a large construction site is an example of a mission-critical communication network. Mission critical communications networks have been used in various sectors such as construction, transportation, utilities, factories, and mining operations (note that in some literature, the term "mission critical communication" is used to refer to the communications systems used by law enforcement and emergency services as well).

2.3.1 Transportation

The transport sector includes buses, trams, trains, railways, subways (aka metros), airports, etc. This sector requires critical communications systems and applications to facilitate critical voice and data communication in promoting safety and efficiency in its operations. The number of transport vehicles, as well as transport alternatives, is increasing to meet the ever-growing populations and their demand for transportation. Therefore, the transport sector needs to have a fast and reliable communications system with voice, data, and video capabilities to support its operations safely and comfortably.

One of the main concerns in transportation operations is the safety of the passengers as well as the employees in transition. The communications system becomes critically important during accidents, especially those involving injuries and criminal activities. Fast and reliable communications between control centers and the employees on the scene provide crucial support in handling these unfortunate events.

The following are some other uses of critical communication systems in the transportation sector:

- Communications with drivers
- Public announcements from control centers
- Intercom systems between drivers and dispatchers
- Transmission of critical data such as vehicle diagnostics, location, and emergencies
- Transmission of data providing navigation information, traffic conditions, video surveillance, etc. for drivers

According to [15], the transportation sector is and will continue to be the largest user of critical communications technologies. For example, the Terrestrial Trunked Radio (TETRA) technology has been used widely in railroads and airports in Europe and in some countries outside Europe [16].

2.3.2 Utilities

The utility sector includes water, gas, telephone, cable, and electricity utilities primarily. The critical communications systems used in this sector typically support service and maintenance teams for their voice and data communications applications.

Transmissions of the readings data from a vast number of utility meters and utility substations dispersed over broad geographical areas seem to be the primary concern in this sector. This type of data tends to be short, needing rather low rates (e.g. 9.6, 16,

32, or at most 64 Kbps). On the other hand, accuracy, reliability, latency, and synchronization are much more stringent [16]. The rather tightly regulated nature of this sector also puts additional pressure to have a much more reliable communication system.

2.3.3 Others

Mission critical systems are used in many other commercial firms and not-for-profit organizations. The following is a list of some examples.

- **Large Campuses** use critical communications systems mainly to support the security staff to maintain the safety and security of the employees and students working and living on the campus. An excellent example of these campuses is large corporate campuses where thousands of employees work. Another good example is educational campuses where thousands of students live in dormitories, attend classes, and participate in curricular and extracurricular activities. Furthermore, a large number of faculty and employees work in classrooms and offices. The security staff relies heavily on critical communications systems to monitor, detect, and respond to any incidents. The security staff collaborates with local public safety agencies to handle many of these incidents. Therefore, the interoperability of campus critical communications systems with the public safety agencies' public safety network is essential.
- **Factories** use critical communications systems for some applications including monitoring the facilities and machinery for incidents (e.g. fires) and security. Stock and inventory management, workers security and safety, as well as workers' job-related communications are some other uses of critical communications systems.
- **Construction Sites** employ critical communications systems to provide communications among construction workers and their supervisor, as well as to monitor the site and control the construction material and various logistical matters.
- The **Oil and Gas Sector** relies heavily on critical communications systems to support its drilling and refinery operations, which are carried out in extremely harsh locations. One primary use of critical communications systems in this sector is the exchange of critical information in real time between thousands of components and control centers (including a collection of technical data from various sensors dispersed at the drilling and refinery sites, and transmitting instructions to all these components to control their operations). An equally important use of critical communications systems is to support the safety of the personnel working at these sites. Since the accidents at these sites may impact the surrounding communities, resulting in environmental disasters, the interoperability of the critical communications systems at these sites with public safety agencies is of paramount importance.

2.4 SUMMARY AND CONCLUSIONS

The chapter provided a discussion on the users of critical communications systems. The discussions included users and organizations providing public safety functions (such as law enforcement agencies and first responders, such as fire departments and ambulatory services) as well as users and organizations in private sectors such as transport, utilities, mining, and similar businesses.

Regardless of the organizations they are deployed in, critical communications systems provide three primary applications: (i) communication among public safety and security personnel to carry out their functions, (ii) communication among the employees (workers) in specific sectors, such as manufacturing and construction, to carry out their job functions, (iii) to collect and exchange technical data from the components and sensors in the field to monitor and control their operations.

REFERENCES

1. "Police," *Wikipedia, the free encyclopedia*. [Online]. Available: https://en.wikipedia.org/wiki/Police. [Accessed: Jul. 29, 2016].
2. "Federal Emergency Management Agency." [Online]. Available: https://www.fema.gov/. [Accessed: May 15, 2018].
3. "United States Coast Guard." [Online]. Available: http://www.gocoastguard.com/. [Accessed: Aug. 2, 2016].
4. "Coast guard," *Wikipedia, the free encyclopedia*. [Online]. Available: https://en.wikipedia .org/wiki/Coast_guard. [Accessed: Jun. 27, 2016].
5. "USCG: Rescue 21." [Online]. Available: https://www.uscg.mil/acquisition/rescue21/. [Accessed: Jul. 30, 2016].
6. M. Zilis, "Gulf coast agencies build network for oil spill," *Radio Resource Mission Critical Communications*, Jun. 16, 2010. [Online]. Available: https://www.rrmediagroup.com/Features/FeaturesDetails/FID/180.
7. "International Federation of Red Cross and Red Crescent Societies," *International Federation of Red Cross and Red Crescent Societies*. [Online]. Available: http://media.ifrc.org/ifrc/. [Accessed: May 15, 2018].
8. American Red Cross, "ARC_3058 Disaster Services Program: Communications," Jul. 1998. [Online]. Available: http://www.qsl.net/ws1sm/ARC_3058.pdf. [Accessed: Sep. 14, 2018].
9. "MOU: Cooperation in emergency telecommunications for disaster preparedness and response." International Federation of Red Cross and Red Crescent Societies and International Amateur Radio Union, 2008. [Online]. Available: http://www.iaru.org/uploads/1/3/0/7/13073366/ifrcandiarumou.pdf.
10. "National Guard of the United States," *Wikipedia, the free encyclopedia*. [Online]. Available: https://en.wikipedia.org/wiki/National_Guard_of_the_United_States. [Accessed: Jul. 14, 2016].
11. "The National Guard—Official website of the National Guard." [Online]. Available: http://www.nationalguard.mil/. [Accessed: Aug. 4, 2016].

12. "Gendarmerie," *Wikipedia, the free encyclopedia*. 10-Jun-2016.

13. "Gendarmerie General Command," *Wikipedia, the free encyclopedia*. [Online]. Available: https://en.wikipedia.org/wiki/Gendarmerie_General_Command. [Accessed: Aug. 2, 2016].

14. "U.S. Army Corps of Engineers." [Online]. Available: http://www.usace.army.mil/. [Accessed: Apr. 30, 2018].

15. J. Gonzalez, "Critical Communications Report," *IHS Markit*, Mar. 21, 2017. [Online]. Available: https://technology.ihs.com/590640/critical-communications-report-2017.

16. TCCA, "Study on the relative merits of TETRA, LTE and other broadband technologies for critical communications markets," Final version 1.1, Feb. 2015.

3

CHARACTERISTICS OF CRITICAL COMMUNICATIONS SYSTEMS

This chapter addresses what users expect from critical communications systems so that they can perform their functions as efficiently as possible. In other words, the chapter focuses on the primary characteristics of critical communications systems, as driven by the needs of users.

3.1 INTRODUCTION

Critical communications systems include a telecommunications network, a set of services and applications, a variety of end-user devices, as well as some operations support systems, also known as network management systems. Critical communications systems also make use of radio frequency bands to exchange the voice, data, and multimedia applications needed to carry out their "critical" functions as well as to transmit and receive information among users in the field and technicians at command centers.

Fundamentals of Public Safety Networks and Critical Communications Systems: Technologies, Deployment, and Management,
First Edition. Mehmet Ulema.
© 2019 by The Institute of Electrical and Electronics Engineers, Inc. Published 2019 by John Wiley & Sons, Inc.

What sets a critical communications network apart from a commercial communications network? Perhaps the most dominant characteristic of critical communications networks is that they provide the basis for *situational awareness* and *command and control* capabilities, which roughly translate into the following capabilities [1]:

- Prioritize delivery of mission-critical data (e.g. bring the dispatch data into the field—ability to send more and detailed information to officers in real time)
- Survive multiple failures (robust, even in extreme conditions; site hardening; enhanced physical protection and battery back-up; redundancy [(intra-network, inter-network; fallback to other networks when needed])
- Maintain data integrity and confidentiality (end-to-end full encryption; link security—both user and control planes; network operations and management security including related data)
- Offer the essential coverage and capacity required (geographical coverage, not population coverage; symmetrical usage [uplink-downlink] pattern, as opposed to downlink heavy commercial pattern)
- Interoperate with other networks and extend coverage and capacity when needed (to enable communication among users outside network coverage, and to secure wide area communication also when users are outside normal network reach)
- Provide right to use and identity management support for officers, applications, and devices (to provision users with "right to use" of resources; dynamic priority and resource management for users and applications)

Note that "national security" and "public safety," are two related, but separate topics. National security is mostly concerned with external/internal threats, whereas public safety concerns include natural disasters, accidents, and deliberately harmful acts.

A variety of factors characterize critical communications systems. These drivers can be grouped into three areas: user needs, technology advances, and network management.

The users of critical communications systems demand specific features and capabilities, which are necessary to perform their tasks. These capabilities may be listed as:

- Features: what the systems should do (e.g. provide group communications)
- Performance: capacity, speed, etc.
- Reliability, survivability
- Security

Another area that plays a primary role in the characterization of critical communications systems is the significant advancement in technologies in a number of related fields such as:

- Communications
- Computing
- Storage
- Security
- Visual (AR, VR)
- Analytics

Finally, the capabilities needed to manage a critical communications system and its operations are drastically different. These capabilities are grouped into two categories:

- Network management (operations, administration, and maintenance of the network and systems)
- Command and control for field operations

3.2 FEATURES COMMON TO BOTH CRITICAL COMMUNICATIONS SYSTEMS AND OTHER WIRELESS NETWORKS

This section provides a list of features that apply to critical communications systems as well as to commercial wireless networks. However, some features are more prominent in public safety networks than in non-public safety networks. Therefore, the features below are described within the context of public safety networks.

- **Authentication:** These features are necessary to manage authorized access to the network and its resources. The system requires public safety personnel to enter some kind of user identification so that the system authenticates and validates the user rapidly and reliably and loads the user's credentials (i.e. profile), which typically include all the resources and features that this specific agent is allowed as previously authorized. For authentication, the system does not rely only on such methods as usernames and passwords. A variety of and a combination of more modern validation technologies such as body part scanning are being incorporated. Another equally essential aspect of authentication is that the equipment integrated into the network is authenticated to make sure that it is authorized to prevent security breaches via maliciously placed equipment within the network.
- **Priority and preemption:** These features allow multiple priority levels and access classes to provide maximum flexibility in assigning initial network

access, radio bearers (packet streams), and in the case of retention and preemption cases, where the level of quality can be decreased for low priority users.

- **Roaming:** This feature allows a user from region A to travel into region B and makes available appropriate features and functionality in the profile of the user in the new region. Additionally, roaming to a commercial network when necessary may also be possible.

- **Anywhere, anytime connections:** These features allow access to connectivity regardless of the location and time. This also implies automatic handover to a commercial network when necessary.

- **Spectrum allocation and sharing:** These features allocate significant spectrum and equipment resources, which is being mandated by the worst-case planning nature of public safety networks (these are stockpiled and unused majority of the time). Allow spectrum sharing, especially during emergencies, to have the ability to accommodate the capacity for public safety when necessary [2].

- **Shared core network:** These features provide a wide range of Internet Protocol (IP) and optical services including IPv4, IPv6, multicasting, Software Defined Networking (SDN) and Network Function Virtualization (NFV), optical wave transport, Layer 2 Ethernet, and at least high definition (HD) quality live streaming.

- **Manageability:** The public safety network must be easy to deploy, integrate, and operate. The network and its components must include state of the art network management protocols, features, and network management systems to administer and maintain the network flexibly and dynamically as well as to keep the network within designed performance boundaries. It is desirable that the network has self-organizing capabilities using autonomic network management techniques to minimize human intervention to maintain and configure the network.

- **Fault management:** These features include the capabilities in critical components (e.g. relays, gateways, cell towers, routing, transmission, and switching equipment) to be able to detect failures (e.g. in power, circuit, software), to locate and repair them as soon as possible. If the repair is not possible in a timely manner, the system is put back into service immediately either by rapid deployment of replacement equipment or temporary deployment of new equipment to keep the network operation going.

- **Self-healing and self organizing:** These features allow the public safety network to autonomously reconfigure itself around failures through the Self-Organizing Network (SON) design.

- **QoS and priority provisioning:** These features allow multiple levels of priority that can be assigned to talk groups and types of calls (e.g. private, group, emergency calls). Incorporate a Quality of Service (QoS) metric that includes priority, guaranteed bit rate, packet delay threshold, and packet error loss rate. Make QoS assignable to each user device.

- **User applications:** These features make it easier for third parties to develop a variety of applications accessing shared services across the public safety system, and allow for real-time sharing of media content.

- **Scalability:** The public safety network can grow or shrink easily depending on future needs. It should be easy to add more users, more equipment, and more capacity when necessary. With the same token, it should be easy to retire equipment and downsize the network as needed.

- **Evolvability:** This feature refers to the capacity of a system for adaptive evolution. The public safety network should be based on well-known standardized technologies, supported by a reputable international community of interest and standard development organizations.

- **Flexible edge:** These features allow nationwide public safety networks to be an interlinked network: mesh and ad-hoc networks at the edge, radio access networks linked to the end users, and a shared core network; all connected via the IP. The mesh or mobile ad-hoc networking capabilities allow the equipment at the edges of the network to serve as packet relays in the formation (in a self-organized way) of a mesh or an ad-hoc network through Interior Gateway Protocols (IPG) and Exterior Gateway Protocols (EPG).

- **Interoperability:** Public safety networks are mostly backward compatible with the previous generations, which are expected to continue to be operational alongside with the new public safety network. Furthermore, some of the public safety networks interoperate with several external networks such as the Internet and commercial wireless mobile networks as well as wired telephony networks. This feature is critical since the external networks mentioned above are frequently used by public safety personnel to communicate with the public or retrieve information that may not be available on public data networks [3].

- **Standards compliance:** The technology, components, interfaces, and protocols used in public safety networks should be based on international standards. Standards permit interoperation of communication devices and systems across a variety of actors in the public safety arena. Technologies that are based on international standards facilitate agencies from multiple countries to collaborate in responding to emergencies thanks to interoperable communications systems. Standardized technology attracts more vendors. This means that there are many competing vendors to choose from, thus driving costs down through economies of scale. Standardized technology allows specialization of vendors; this means that the network may be composed of a number of different components acquired from a variety of different vendors, interworking together thanks to standardized interfaces and protocols.

3.3 FEATURES UNIQUE TO CRITICAL COMMUNICATIONS SYSTEMS

This section provides a list of features that mostly apply to critical communications systems.

- **User-specified unique features:** Perhaps the essential characteristic of the public safety network is that a significant majority of features have been defined by the users (i.e. the agencies and stakeholders involved in public safety). Additionally, these features are provided reliably, securely, and with a latency expected by the users. Note that public safety networks have traditionally been designed based on user requirements at the "worst case" levels that are necessary during emergency (disasters) situations. It is not assumed that the network will not need these levels during normal operations.

- **Performance:** The usage patterns on public safety networks are unpredictable. No one knows when or where an emergency might occur. Therefore, public safety networks are designed and engineered to support sudden, unexpected spikes in usage in any portion of the network. Throughput, capacity, and latency in public safety systems are specified, designed, engineered, and built for worst-case scenarios, unlike commercial networks, which are typically engineered for best effort. The network is optimized for peak performance for a relatively small number of extremely demanding users (compared to commercial networks).

- **Broadband data rates:** This feature applies to the Long Term Evolution (LTE)-based public safety networks. The network supports broadband data rates to allow public safety agencies to exchange high definition videos, graphics, photos, and high fidelity audio, etc., in real time. Streaming video from major incidents to a central control provides valuable information in determining the best way to respond. Also, the ability to pass medical telemetry as well as live video of the patient can be a lifesaver.

- **Real-time access to a vast amount of information:** Allow queries to be delivered almost in real time in information-rich, appropriately formatted reports with high-resolution images, high fidelity audio clips, and HD quality video clips.

- **Availability/Serviceability:** The public safety network is built to guarantee the highest level of service, so that first responders never find themselves without the ability to communicate. QoS and Quality of Experience (QoE) meet at the levels approved by all public safety users.

- **Reliability/Resiliency:** Public safety networks are highly reliable; built to withstand (survive) any emergency, such as earthquakes, hurricanes, tornados, forest fires, floods, and landslides. To ensure maximum robustness, all portions of the network are typically fully redundant, including the fiber backbone to support the wireless portion of the network. Rapid deployment of temporary infrastructure is an essential aspect of this feature.

- **Security:** Public safety networks can encrypt both user and control data with a relatively large encryption key. Application layer encryption is part of the end-to-end security feature as well.

- **Location awareness:** Public safety networks, together with sensors and Global Positioning System (GPS) devices in the network, help first responders to collect relevant data including GPS coordinates and location information based on radio triangulation techniques.

- **Situational awareness:** These features allow a significant amount of essential data to be exchanged between back office applications and public safety officials on the scene to analyze the situation and collaborate with others.

- **Direct mode communications:** These features allow the users to have device-to-device direct communications so that public safety officials can communicate with each other when there is no coverage through the infrastructure.

- **Group (multicasting and broadcasting) communication:** These features facilitate the connection of a large number of users (hundreds of them) in a local area or geographically dispersed locations in a timely way. Public safety agencies often work in groups and need to communicate in groups. This is a must-have feature.

- **Over the air rekeying (OTAR):** These features allow end-user devices to be able to obtain and change or update the keys used in encryption over the air (this minimizes the physical/manual distribution of the encryption keys, thus minimizing the risk of keys stolen or lost).

- **On-scene, in-field capabilities:** These features allow the personnel to establish and maintain networks on-scene locally (e.g. in a building) quickly and transparently in emergency situations to allow the on-scene personnel to exchange data.

- **User devices:** In critical communications systems, the personnel uses a diverse set of interoperable end devices with various modes of operation (Land Mobile Radio [LMR], LTE, 3G, 4G, Wi-Fi, etc.). Also, the devices need to have the appropriate ergonomics and ruggedness required during public safety operations. Some additional important features of devices include hands-free or with one hand operations with protective gear in place including gloves, and the protection (secure, stable, portable, etc.) of devices used in vehicles.

- **Battery life:** These features allow mobile wireless devices to have a battery life that exceeds the existing systems. Battery life features typically include a duty cycle combination of transmission, speak, and standby.

3.4 IMPORTANCE OF INTEROPERABILITY FEATURES

The word interoperability is a loaded one. Its most comprehensive definition includes "governance, standard operating procedures, technology, training/exercises, and usage of interoperable communications" [4]. From communications aspects, the word is used to mean that, for a given standard technology, the components built by different manufacturers work together. For example, an agency building a Project 25 based network acquires equipment from vendor X and vendor Y. The agency would want some guarantee that this equipment provided by two different vendors works when connected. This is typically verified by a set of *conformance* tests. All the technologies mentioned in this book have a set of well-defined procedures and standards to obtain *certificates* to prove the interworking of the equipment built by different vendors.

The same word, "interoperability," is also used to mean that different networks owned by different agencies work together. For example, an agent on network A should be able to communicate with another agent on network B. Network A and network B could be based on the same technology or each may be based on a different technology.

Interoperability is a must feature in critical communication systems. Unfortunately, interoperability is one of the weakest links in current critical communications, especially in the public safety area [5–7].

In several countries, there are no centralized common public safety networks that all agencies can share. It is most likely that different agencies use different communications technologies (interoperability problems may still be present due to differences in implementation, operation, and even jurisdiction).

Natural and manmade disasters have shown us that all agencies cooperating during such disasters must be able to communicate to help the public. Therefore, interoperability among all the networks (regardless of the technologies) used by all agencies is a paramount interest.

Currently, temporary arrangements are used for interoperability between two or more incompatible radio systems (e.g. "analog patching" between networks). Proprietary solutions also include interoperability via gateways, which use the same protocol for translating voice and data. This facilitates radios and protocols with different technologies to communicate.

There are a bunch of interfaces and capabilities required for each technology to interwork with other technologies (this should include the networks based on analog technologies, which may be around for a while) (Table 3.1).

Also, applications, administration, operations, and security systems of each network should be configured to interoperate. Furthermore, public safety agencies may use commercial landline and mobile networks as well as Wi-Fi networks, especially during emergencies. Therefore, interoperability scenarios should also include these types of networks [8].

TABLE 3.1. An Illustration of Possible Interoperability Scenarios

	Analog	P25	TETRA	DMR	LTE
Analog	x	x	x	x	
P25	x	x	x	x	x
TETRA	x	x	x	x	x
DMR	x	x	x	x	x
LTE		x	x	x	x

The Project 25 based system is already backward compatible with the existing Digital Mobile Radio (DMR) and other analog systems [4]. Furthermore, interoperability with commercial systems is also essential especially during emergencies. Since the emergency call number system is one of the primary triggers for public safety activities, it is vital that public safety networks must be interoperable with emergency call centers as well.

There are several vendors offering solutions to provide complete interoperability with LTE-based and Project 25 based systems [9]. Since the intersystems interface of Project 25 is based on the IP/Transmission Control Protocol (TCP) standards including Session Initiation Protocol (SIP) and Real-time Transport Protocol (RTP), the interoperability between these two should be relatively straightforward [10]. These are also included in the LTE standards.

Next generation technology independent interoperability of emergency services (GERYON) was a European Union (EU) project. Its objective was to integrate the communication networks used by emergency and safety management bodies—ambulances, fire brigades, civil protection teams—with new generation telephone networks (4G, LTE). The project work plan defined a series of design and implementation work packages aimed at developing a noncommercial demonstrator prototype. At the end of the project, all its objectives had been successfully fulfilled, resulting in a fully working IMS compatible ecosystem capable of providing today Private Mobile Radio (PMR) grade communications while paving the way for future professional LTE networks [3].

3.5 SUMMARY AND CONCLUSIONS

This chapter presented some of the most salient characteristics of critical communications systems. There are a variety of factors that characterize critical communications systems. These factors can be grouped into three areas: user needs, technology advances, and network management. Some of these features are unique; some are also applicable to other wireless systems such as commercial cellular networks. However, the most dominant characteristic of critical communications networks is that they provide the basis for *situational awareness* and *command and control* capabilities.

Some of the features common to all wireless networks include authentication, anywhere-anytime connections, spectrum allocation and sharing, manageability, QOS and priority provisioning, scalability, evolvability, and interoperability.

Some of the features that are unique to critical communications systems include user-specified unique features, performance, real-time access to vast amounts of information, availability, serviceability, resiliency, security, location awareness, situational awareness, direct mode communications, group (multicasting and broadcasting) communication, OTAR, on-scene in-field capabilities, unique user devices, and battery life.

The interoperability feature is of paramount importance to public safety agencies. Unfortunately, interoperability is one of the weakest links in current critical communications, especially in the public safety area. Currently, temporary arrangements and proprietary solutions are used for interoperability between two or more incompatible radio systems. However, interoperability is a must-have feature in critical communication systems. Therefore, continuous effort by the standards bodies in this area is needed, especially now that LTE-based public safety networks need to work side by side with older technologies for a long while.

REFERENCES

1. National Public Safety Telecommunications Council, "Defining public safety grade systems and facilities final report," An NPSTC Public Safety Communications Report, May 2014.
2. "Public Safety Spectrum Trust." [Online]. Available: http://www.npstc.org/psst.jsp. [Accessed: Sep. 14, 2018].
3. "GERYON (Next Generation Technology Independent Interoperability of Emergency Services)," EU CORDIS, 284863.
4. B. Deverall, "Briefing for e-gif working group on incorporating the APCO P25 standards into the e-government interoperability framework," Added Value Applications, Sep. 19, 2006.
5. W. Lehr and N. Jesuale, "Spectrum pooling for next generation public safety radio systems," in *IEEE Proc. Dynamic Spectr. Access Netw. (DySPAN2008) Conf.*, Chicago, Oct. 2008, pp. 1–23.
6. P. R. Kempkerhttp, Department of Homeland Security, "Basic gateway overview." [Online]. Available: http://www.c-at.com/Customer_files/Fairfax%20rally/Module%201% 20-%20Basic%20Gateway%20Overview.ppt. [Accessed: Sep. 13, 2018].
7. TETRA and Critical Communications Association, "Broadband spectrum for mission critical communication needed," Position Paper, Aug. 2013.
8. Tait Limited, "Introducing Unified Critical Communications," White Paper, 2014.
9. D. Witkowski, "Effectively testing 700 MHz public safety LTE broadband and P25 narrowband networks," White Paper, Anritsu Company, 2013. Available: https://dl.cdn-anritsu .com/en-us/test-measurement/files/Technical-Notes/White-Paper/11410-00601B.pdf
10. D. Tuite, "Can public-safety radio's P25 survive LTE?," *Electronic Design*, Jul. 17, 2012. [Online]. Available: https://www.electronicdesign.com/analog/can-public-safety-radio-s-p25-survive-lte.

4

INTRODUCTION TO TECHNOLOGIES AND STANDARDS FOR CRITICAL COMMUNICATIONS

This chapter provides an overview of the technologies, protocols, and related standards used in building and operating critical communications systems. After a brief introduction, details of each technology are provided in the following chapters.

4.1 INTRODUCTION

As stated before, critical communications systems consist of wireless and wired communication networks as well as a variety of application and supporting systems. Voice and data have been two main types of communications historically. Now, pictures and video have also been transmitted in new generation critical communication systems.

Information stored, transmitted, and processed by critical communications systems can be in analog or digital form. Similarly, information can be transmitted using either analog or digital signals. Various conversion techniques such as modulation and coding are used if there is a mismatch between the format of the information and the type of transmission signal.

Fundamentals of Public Safety Networks and Critical Communications Systems: Technologies, Deployment, and Management,
First Edition. Mehmet Ulema.
© 2019 by The Institute of Electrical and Electronics Engineers, Inc. Published 2019 by John Wiley & Sons, Inc.

Like communication technology, critical communication systems, collectively referred to as Land Mobile Radio (LMR) or Private Mobile Radio (PMR) systems, have evolved. For a long time, analog radio-based technology has dominated critical communication systems. Even today, a surprisingly significant number of agencies around the world continue to use analog radio systems, which support mainly voice communication.

Old analog critical communications radio technologies are being replaced in most of the world by all-digital technologies based on narrowband and broadband transmission capabilities.

There are mainly three narrowband digital radio technologies in use today. These are Project 25, Terrestrial Trunked Radio (TETRA), and Digital Mobile Radio (DMR) standards.

All these narrowband technologies support a wide range of voice and a somewhat limited set of data applications. Project 25 is more popular in North America, whereas TETRA is more popular in Europe and around the world. DMR is a relatively newer narrowband technology, which is used in some regions.

The architecture of an LMR/PMR system can be designed to form either a conventional or a trunked system. A **conventional system** is typically a simple system used for smaller groups, in relatively smaller geographical areas (a repeater supporting around less than 100 radios). It is a repeater radio system used to repeat calls among channels, with taller sites and high power. Radio channels are selected by the user since no control channel or core switching is used in conventional systems. In other words, the operation of the system is controlled by the radio user. Figure 4.1 shows a typical, simplified configuration of a conventional radio system.

A **trunked system** is typically used for larger groups and broader geographical areas. Trunked systems incorporate a controller with a control channel and one or more switches in its infrastructure. In a trunked system, channels are assigned automatically to users. In other words, the operations of the system are controlled automatically. See Figure 4.2 for a typical trunked system configuration.

Table 4.1 provides a summary of the critical aspects of conventional and trunked radio systems.

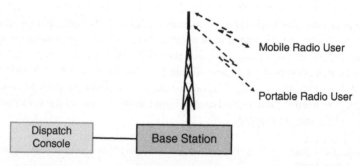

Figure 4.1. Typical configuration of a conventional radio system.

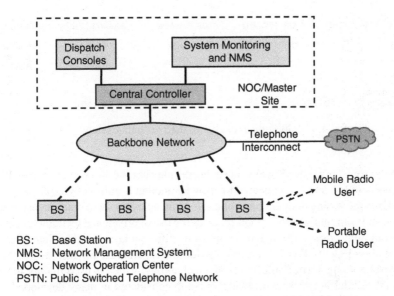

BS: Base Station
NMS: Network Management System
NOC: Network Operation Center
PSTN: Public Switched Telephone Network

Figure 4.2. Typical configuration of a trunked radio system [1].

LMR and PMR systems typically operate in the Very High Frequency (VHF) and Ultra High Frequency (UHF) bands. The specific frequency bands vary from region to region. Figure 4.3 shows typical frequency ranges used in the USA and internationally for narrowband analog and digital critical communication systems.

To accommodate emerging broadband critical communication systems, regulators around the world allocate new frequency bands. For example, 10 MHz wide for each direction at 700 MHz for public safety applications has been allocated in the USA.

PMR/LMR-based technology (analog and narrowband digital) has been the primary means of instant communications for many commercial firms and public safety

TABLE 4.1. Comparison of Conventional and Trunked Systems [1]

Conventional	Trunked
Radio channels selected by the user	Radio channels assigned to the user automatically
Channel access instantaneous	Channel access time varies with technology and other factors (typically measured in hundreds of milliseconds)
No control channel needed	Control channel used in all trunked applications
Suitable for smaller groups of users	Better for larger (~300+ users) organizations
Switching not needed	Core switching essential for the operation
Digital control signaling inherent	Digital control signaling inherent

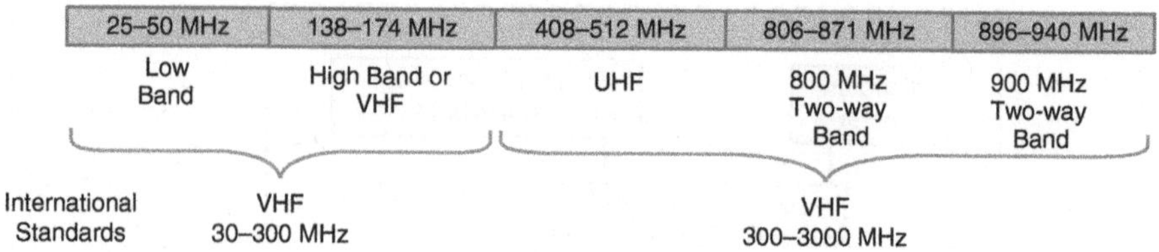

25–50 MHz	138–174 MHz	408–512 MHz	806–871 MHz	896–940 MHz
Low Band	High Band or VHF	UHF	800 MHz Two-way Band	900 MHz Two-way Band

International   VHF   VHF
Standards   30–300 MHz   300–3000 MHz

Figure 4.3. Traditional radio bands used for public safety systems [2].

agencies for a long time. The existing subscription base is estimated to be over 40 million users today, and it is expected to grow by another 5 million by 2017 [3].

Ever-increasing demand by public safety agencies for more data-intensive application and communication support requires the use of broadband technologies in critical communication. Although there are some different broadband technologies such as Wireless Fidelity (Wi-Fi) and Worldwide Interoperability for Microwave Access (WiMAX), Long Term Evolution (LTE) technology, which is used mainly for commercial cellular networks worldwide, is being considered in many developed countries, including the USA and Europe, for its ubiquitous availability.

4.2 ANALOG SYSTEMS—HISTORICAL PERSPECTIVE

The use of analog technology goes back more than 100 years. One-way Amplitude Modulation (AM) radio based commercial broadcasts in the 1930s were often used to send emergency messages to law enforcement agencies. Two-way AM broadcast for use by police was introduced in the 1930s, followed by the introduction of Frequency Modulation (FM) radio technologies in the 1940s. Hand-carried radios used in vehicles of public safety officials were introduced in the 1950s. Evolving from two-way radios, critical communication systems where Radio Frequency (RF) channels were shared by many users were built in the 1960s [4]. In these systems, information (mainly voice) and signaling are transmitted in the analog format.

Following the advances in information and communication technology from analog to digital, critical communication systems joined the trend and in the 1970s and 1980s, some digital technology-based systems began to emerge. There are still many agencies around the world that use analog critical communications systems. As shown in Figure 1.2, analog systems still constitute the majority of communications today, although it is diminishing slowly.

This 100-year-old analog technology outperforms current state of the art digital systems in voice quality (intelligibility) in point-to-point communications.

One of the first efforts to standardize an analog-based critical communication network took place in the late 1970s by the Association of Public Safety Communications Officials (APCO). The resulting standard, APCO-16 (or Project-16),

TABLE 4.2. A list of key features of APCO-16 [2]

Specifications	APCO-16
Technology	Analog
Multiple access technique	FDMA
Telephony type	Trunked
Channel bandwidth	25/30 KHz
Frequency range	VHF (136–174 MHz), UHF (403–512 MHz), and 800 MHz bands

defined a framework and operational requirements for a trunked analog radio system. However, the implementation of Project-16 was not successful due to costly and proprietary implementation of different vendors, resulting in incompatible products. Table 4.2 lists several important specifications of APCO-16 [5].

4.3 NARROWBAND LAND AND PRIVATE MOBILE RADIO SYSTEMS

Digital narrowband radio systems provide facilities for closed user groups, group call, and push-to-talk, and have call set-up times that are short compared to those of cellular systems by using digital communications techniques instead of analog techniques. Many PMR systems allow Direct Mode Operation (DMO), in which terminals can communicate with one another directly when they are out of the coverage area of the network.

PMR radio equipment may be based on such standards as Ministry of Posts and Telegraph (MPT)-1327 [6], TETRA [7], TETRAPOL [8], Project 25 [9], digital PMR (dPMR) [10], and DMR [11]. Some of the PMR equipment is designed for dedicated use by specific organizations, or standards such as integrated Digital Enhanced Network (iDEN) (aka Digital Integrated Mobile Radio Service [DIMRS]) [12], Next Generation Digital Narrowband (NXDN) [13], Integrated Dispatch Radio (IDRA) [14], Frequency Hopping Multiple Access system (FHMA) [15], and Global Open Trunking Architecture (GoTa) [15], intended for general commercial and other use.

Project 25 is a standard developed by the Telecommunications Industry Association (TIA) as TIA-102 Project 25. TETRA is a standard developed by the European Telecommunications Standards Institute (ETSI). TETRAPOL is a digital PMR standard developed by another industry group in Europe [8]. TETRAPOL is not related to TETRA, other than the use of the acronym "TETRA," as discussed briefly in Chapter 1. DMR is another ETSI standard (ETSI DMR TS102 361). dPMR technology is another ETSI standard that has evolved to provide equivalent FDMA standards for DMR Tier II and Tier III. The NXDN protocol was developed jointly by Icom Incorporated and Kenwood Corporation.

As discussed before, Project 25, TETRA, and DMR have been the dominant narrowband systems used by public safety agencies in the world today. Therefore, the following chapters provide detailed discussions about only these three technologies. See Table 1.1 for a comparison of salient features of these three standards.

4.4 LIMITATIONS OF NARROWBAND PMR/LMR SYSTEMS

Although narrowband technologies provide excellent voice communications related features, they are incapable of providing support for handling a significant amount of data. The problem is the very narrow bandwidth of the channels, which restricts data rates. Narrowband PMR/LMR channels are designed to support only around 9.6 Kbps to 36 Kbps. These rates are somewhat slow and not sufficient compared to what commercial systems' Mbps and Gbps offer. Even though several relatively higher data rate enhancements have been introduced, such as TETRA Enhanced Data Service (TEDS) for TETRA, it is not enough to handle today's data-intensive applications, which may include real-time high-resolution image and video transmission.

Lower indoor and rural handheld coverage is another design and economic limitation of narrowband systems today.

Costs of acquiring, deploying, and managing narrowband LMR systems are rather high since the potential target market for these systems is rather small, not taking advantage of the economy of scale.

The narrowband LMR systems in use today are fragmented and support limited interoperability. Interoperability is perhaps the most significant issue considering the rise of terrorism worldwide and the demand for the highest level of interoperability among all agencies involved, especially in emergency situations.

Today, many countries are looking into the viability of a nationwide standard, interoperable, broadband mobile wireless network that will allow all public safety agencies to use the same network and provide higher data rates to support data-intensive applications.

4.5 BROADBAND TECHNOLOGIES

The word *broadband* may mean different things to different people. In the communications and networking field, the term broadband refers to a communications technology that uses a relatively larger bandwidth, thus offering relatively higher data rates. There is no industry-wide consensus as to the boundary between narrowband and broadband data rates. However, 1 Mbps seems to be a good choice, as suggested by many. However, in this book, we use this term to refer to the public safety systems providing high-speed data transmission rates well above those provided by the narrowband LMR systems discussed in the previous section.

Some widely used technologies may qualify as "broadband." These include Wi-Fi, WiMAX [16], and LTE. Both Wi-Fi and WiMAX are standardized by the Institute of Electrical and Electronics Engineers (IEEE). The 3rd Generation Partnership Project (3GPP) standardizes LTE (see Figure 1.3 for a comparison of two important aspects of networking for some technologies).

As discussed before, LTE and LTE-Advanced (LTE-A), the next version of LTE, are the only accepted technology worldwide as 4th Generation (4G) mobile broadband communications systems. LTE-A, theoretically, supports up to 100 MHz bandwidth and 1–3 Gbps (downlink) peak data rate. LTE and LTE-A based commercial networks are widely deployed all around the world.

LTE (and LTE-A) is an attractive broadband technology for building next generation critical communications systems, compared to others. A well-proven technology, relatively overall low cost due to the economy of scale, high data rates, broad coverage areas, and ubiquitous availability make LTE technology a clear choice. However, some essential features required in critical communications are missing in LTE technology. Lack of mission-critical voice applications such as push-to-talk and group calling and DMOs are some of the leading concerns. Currently, 3GPP and other stakeholders have been working on incorporating these features into the upcoming LTE specifications.

As mentioned above, the WiMAX technology specified in IEEE 802.16 also provides wide area broadband communications with similar broadband capabilities as LTE technology. It competed with LTE as a candidate for 4G commercial cellular networks for a while. In the end, LTE became the choice for 4G for that market. However, the WiMAX Forum, which is an industry consortium promoting the use of WiMAX in various areas, has some initiatives that suggest the use of WiMAX in mission-critical communications [16]. Airport surface communications systems, smart grid, utility private networks, and oil and gas networks are some of these initiatives [17].

4.6 OTHER TECHNOLOGIES

The technologies discussed above dominate critical communications systems; they play critical roles in defining the overall system. In other words, the systems are identified with these technologies. However, some other Information and Communication Technologies (ICT) are used by agencies; perhaps some are in noncritical areas.

Commercial satellite systems with voice and low data rates have been used by public safety agencies to provide support in several areas including fixed locations connectivity, deployable fixed operations, and fully mobile operations, especially in rural and remote areas [18]. Satellite systems can be highly useful as a temporary or backup communication system during emergencies such as disasters.

Wired and wireless Local Area Networks (LANs) connect office computers, laptops, tablets, printers, servers, and other office equipment as well as applications and support systems. The Ethernet technology specified in IEEE 802.3 dominates the wired LAN market, while the Wi-Fi technology specified in the IEEE 802.11 series dominates the wireless LAN market.

Wi-Fi technology may be used in the field as well either by using an existing commercial "hot spot" or a temporarily set up Wi-Fi network during emergency situations.

The Bluetooth technology specified in IEEE 802.15 can be used by agents in the field to provide wireless voice communication between headsets and terminal devices (portable radios). Furthermore, Bluetooth is a Personal Area Network (PAN) technology that can be used to provide secure wireless data and video communications among various devices around a field agent. These devices may include backpack-transported computers, video cameras, wearable keyboard and wireless mouse, eyeglass-mounted visual displays, and others.

Amateur radio (aka ham radio) technologies are also used in public safety in emergency situations. Ham radio operators, hobbyists, quite often get involved in helping first responders in restoring emergency communication. For example, during Hurricane Katrina, many amateur ham operators, from across the country, traveled to disaster areas where commercial and critical communication infrastructure were knocked down, bringing their radios, antennas, and amplifiers, setting up communications for first responders [19].

Hams use a variety of voice, text, image, and data communications technologies in the analog or digital format. Encryption is not permitted. Modes of communication include both modulation types (AM, FM, Phase Modulation [PM]) and operating protocols (packet radio, digital mobile radio, PACTOR) [20].

Another wireless technology that may be used by public safety agencies, not necessarily in emergency situations, is the Radio Frequency ID (RFID) technology. RFID together with other technologies is and can be used much needed automated and efficient management of the registry and check out processes of various assets such as vehicles, weapons, radar guns, and other assets that agents use in the field [21]. RFID technology may be either passive or active using Ultra High Frequency (UHF) channels. Tags attached to assets communicate with typically handheld readers at a distance ranging from 2 to 30 feet. The readers, in turn, communicate with back office servers with appropriate applications to transmit the collected data for further processing.

Finally, the Internet is a technology widely used by public safety agencies to gather and transmit information, to communicate with a variety of application servers, as well as to take advantage of "cloud" based services. The cloud includes some services typically offered by third parties to store and access data, software, and even the virtualized infrastructure through the Internet.

Public safety agencies can take advantage of cloud services such as Infrastructure-As-A-Service (IAAS) or Software-As-A-Service (SAAS) to outsource their technology to the cloud. This move will help public safety agencies to ease the management of their ICT, to provide a more secure environment, and to focus on the analysis of information, rather than operating the infrastructure [22].

4.7 SUMMARY AND CONCLUSIONS

Technologies used in critical communications systems include several decades old analog systems mainly for voice; two decades old narrowband all digital radio-based systems (like Project 25 and TETRA) for voice and low bit rate data; and more recent broadband mobile wireless systems like LTE for higher data rates to support data-intensive multimedia applications.

Old analog technologies are being phased out slowly to make room for all digital narrowband technologies. There are also serious discussions and plans in some countries to use LTE technology for critical communications systems.

As of writing this book, intensive research and development efforts are underway to develop so-called 5th generation (5G) mobile cellular technologies. 5G is expected to provide much higher data rates and much lower latency. Additionally, 5G is envisioned to incorporate other technologies including small cells (Wi-Fi), the IoT, augmented reality, and others.

REFERENCES

1. Tait Communications, "P25 Best Practice presented by Tait Communications," P25 Best Practice, Jan. 23, 2014. [Online]. Available: http://www.p25bestpractice.com. [Accessed: Jan. 10, 2017].
2. "Public Safety Technical Assistance Tools Website." [Online]. Available: http://www.publicsafetytools.info/training/training_radio_101_info.php. [Accessed: Jan. 14, 2017].
3. Signals and Systems Telecom, "The public safety LTE & mobile broadband market: 2012–2016," Market Report, Nov. 2012.
4. NPSTC, "LMR 101 NPSTC Final." [Online]. Available: www.npstc.org/documents/LMR%20101%20NPSTC%20Final.ppt. [Accessed: Sep. 13, 2018].
5. RF Wireless World, "APCO-25 vs APCO-16 | Difference between APCO-25 and APCO-16." [Online]. Available: http://www.rfwireless-world.com/Terminology/APCO25-vs-APCO16.html. [Accessed: Jan. 3, 2017].
6. "MPT-1327." [Online]. Wikipedia. Available: https://en.wikipedia.org/wiki/MPT-1327 [Accessed: Jun. 29, 2016].
7. J. Dunlop, D. Girma, and J. Irvine, *Digital Mobile Communications and the TETRA System*. John Wiley & Sons, 2013.
8. "Home | Tetrapol Forum." [Online]. Available: http://www.tetrapol.com/. [Accessed: Jul. 24, 2016].
9. "Project 25 Technology Interest Group." [Online]. Available: http://www.project25.org/index.php. [Accessed: Jan. 10, 2017].

10. "Welcome to the dPMR Association, Narrowband 6.25Khz FDMA Digital." [Online]. Available: http://www.dpmr-mou.org/. [Accessed: Jan. 21, 2017].
11. Digital Mobile Radio Association, "The DMR Standard," Dec. 29, 2010. [Online]. Available: http://dmrassociation.org/the-dmr-standard/. [Accessed: Jul. 24, 2016].
12. "iDEN." [Online]. Wikipedia. Available: https://en.wikipedia.org/wiki/IDEN. [Accessed: Dec. 31, 2016].
13. "NXDN™: A brief overview I What is NXDN™? I NXDN Forum Website." [Online]. Available: http://www.nxdn-forum.com/what-is-nxdn/nxdn-a-brief-overview/. [Accessed: Jan. 17, 2017].
14. "Integrated Dispatch Radio System." ARIB Standard, RCR STD-32A. Nov. 15, 1995.
15. "Digital land mobile systems for dispatch traffic," ITU-R WP 5A, 5A/TEMP/14, Jun. 2012.
16. WiMAX Forum, "WiMAX Forum I AeroMACS, WiGRID, and WiMAX Advanced Technologies." [Online]. Available: http://www.wimaxforum.org. [Accessed: Jul. 24, 2016].
17. WiMAX Forum, "WiMAX Forum initiatives: AeroMACS, WiGRID, WiMAX Advanced & others." [Online]. Available: http://www.wimaxforum.org/Page/Initiatives. [Accessed: Jan. 6, 2017].
18. APCO International, "Brief history overview of broadband for public safety." [Online]. Available: https://www.apcointl.org/doc/spectrum-management/18-brief-history-overview -of-broadband-for-public-safety/file.html. [Accessed: Sep. 13, 2018].
19. "Ham radio volunteers worry about spectrum plan." [Online]. Available: http://www.npr .org/2011/04/30/135873302/ham-radio-volunteers-worry-about-spectrum-plan. [Accessed: Jan. 4, 2017].
20. "Amateur radio." [Online]. Wikipedia. Available: https://en.wikipedia.org/wiki/Amateur_ radio. [Accessed: Jan. 4, 2017].
21. Omnitrol Networks, "Asset tracking for public safety RFID & wireless technologies for efficient asset management." [Online]. Available: http://www.falkensecurenetworks.com/ PDFs/Asset_Tracking_for_Public_Safety.pdf. [Accessed: Sep. 13, 2018].
22. "Why take public safety data into the cloud?" [Online]. Available: http://www.policemag .com/channel/technology/articles/2015/08/why-take-public-safety-data-into-the-cloud .aspx. [Accessed: Jan. 7, 2017].

5

PROJECT 25 (P25)

This chapter presents Project 25, a narrowband Land Mobile Radio (LMR) technology that provides voice and low-speed data communication capabilities for critical communication systems. The chapter discusses in high-level Project 25's technical specifications, network architecture, interfaces and protocols, and services. The standardization process and its future are also discussed. Appendix A provides a comprehensive list of Project 25 related standard documents.

5.1 INTRODUCTION

P25 stands for Project 25, which is the project name and number given by the Association of Public-Safety Communications Officials (APCO) to this project to develop a public safety digital LMR standard; that is why it is sometimes called APCO-25.

In 1988, the following US federal agencies established an 11-member steering committee, called the Project 25 Steering Committee, to define the requirements

Fundamentals of Public Safety Networks and Critical Communications Systems: Technologies, Deployment, and Management, First Edition. Mehmet Ulema.

© 2019 by The Institute of Electrical and Electronics Engineers, Inc. Published 2019 by John Wiley & Sons, Inc.

for a public safety digital radio communication system to be known as Project 25 [1]:

- Association of Public-Safety Communications Officials—International (APCO)
- National Association of State Telecommunications Directors (NASTD)
- National Telecommunications and Information Administration (NTIA)
- National Communications System (NCS)
- National Security Agency (NSA)
- Department of Defense (DoD)

Also involved in this effort were the Department of Homeland Security Federal Partnership for Interoperable Communication (FPIC), the Coast Guard, and the National Institute of Standards and Technology (NIST).

The Project 25 Steering Committee established an arrangement with the Telecommunications Industry Association (TIA) for a joint effort as follows:

- Project 25 Steering Committee, representing the user community, develops requirements, which are formally documented and called the Statement of Requirements (SoR).
- TIA, having established a reputation for developing well-known telecommunication standards especially in the wireless mobile communications systems area, develops standardized specifications based on the requirements provided by the user community.

Project 25 has been accepted as a national standard in the US for critical communications systems serving local, state, county, and federal public safety-related agencies. The Project 25 suite of standards is designated as TIA-102 [2, 3].

Both user requirements and the resulting standards specifications are continually updated to accommodate new user requirements, new regulations, and new technologies, as well as to incorporate experiential improvements. Accordingly, Project 25 features and capabilities are typically discussed under the following three categories, called phases:

- Phase 0 refers to legacy/proprietary (i.e. non-Project 25) requirements and standards for an analog air interface and the supporting legacy system (i.e. radios and infrastructure).
- Phase 1 refers to Project 25 requirements and standards for a digital common air interface (CAI) (based on Frequency Division Multiple Access [FDMA] using a 12.5 KHz channel) and for the supporting system (i.e. radios and infrastructure). Phase 1 equipment operates in analog, digital, or mixed mode for

backward compatibility. It supports 4800 symbols per second (two bits of data for a raw bit rate of 9600 bps). It uses FDMA, which consists of a Continuous 4 level FM (C4FM) modulated signal or a Compatible Quadrature Phase Shift Keying (CQPSK) modulated signal in a 12.5 kHz channel (subscriber equipment transmits in C4FM; the site equipment may transmit at C4FM or CQPSK). Some other Phase 1 features can be listed as follows:

– Project 25 trunking control channel (all features)

– Project 25 trunking traffic channel (all features)

   • Supports data transmission, either piggybacked with voice, or up to 9.6 Kbps on the traffic channel

– Project 25 conventional channel (all features)

– Analog FM (plus autosense between analog and digital)

– Simulcast

– Fallback to conventional operation when not connected to Base Station Controller (BSC).

• Phase 2 refers to Project 25 requirements and standards for a digital CAI and the supporting system (i.e. radios and infrastructure). CAI is based on a two-slot Time Division Multiple Access (TDMA) signal and uses a 6.25 KHz channel or an equivalent bandwidth. It supports 6000 symbols per second where each symbol encodes two bits of data for a raw bit rate of 12,000 bps.

   Fixed site output modulation is Harmonized Differential Quadrature Phase Shift Keying (H-DQPSK) with subscriber units using Harmonized Continuous Phase Modulation (H-CPM) on the input.

   Note that the TDMA CAI in Phase 2 is an addition to the standard; it does not replace the FDMA CAI. This addition allows existing 12.5 KHz wide license holders to double call capacity by upgrading their infrastructure to Phase 2. Some other Phase 2 features are:

– two-slot TDMA traffic channel

– H-CPM uplink, H-DQPSK downlink

– Mode management with Phase 1

– Call continuation

– Traffic to control channel synchronization (option)

– Audio and call preemption (option)

– Closed loop power control (option)

– Full duplex (option)

The vocoder in Project 25 radios produces a digital bit stream that is relatively easy to encrypt.

Table 5.1 provides a comparison of several key features of Phase 1 and Phase 2 of Project 25 standards.

TABLE 5.1. A Comparison of Project 25 Phases

Functionality	Phase 1	Phase 2 (FDMA)	Phase 2 (TDMA)
Standards organization	TIA	TIA	TIA
Channel access method	FDMA	FDMA	TDMA
Channel bandwidth	12.5 KHz	12.5 KHz	6.25 KHz
Modulation	C4FM (or CQPSK linear modulation	C4FM (or CQPSK linear modulation)	H-DQPSK (linear modulation)
Raw data rate	9.6 Kbps	9.6 Kbps	12 Kbps
Number of time slots	N/A	N/A	2
Direct mode	Yes	Yes	Yes
Trunking mode	Yes	Yes	Yes
Analog fallback	Yes	Yes	Yes
Speech Codec	IMBE, AMBE	IMBE, AMBE	AMBE (Half rate)

One of the early focuses was interoperability. The steering committee wanted Project 25 to provide a standardized common digital radio communication system so that all agencies involved in public safety could communicate, especially during emergency situations. An agreement was reached with TIA's TR-8 engineering standard committee to develop a set of standards specifications based on the requirements established by the steering committee [1]. This joint effort resulted in a series of standards numbered as the TIA-102 series [3]. Appendix A provides a complete list of the TIA standards developed for Project 25.

The resulting Project 25 specifications are based on an open architecture with standardized, publicly available, open interfaces, features, and operations. The primary objective is to allow interoperability among Project 25 compliant equipment, as well as interoperability with the legacy radios operating in analog mode. In other words, Project 25 radios can operate in both analog and digital modes:

- Project 25 radio–analog radio (analog mode)
- Project 25 radio–Project 25 radio (analog or digital mode)

The Project 25 series of standards supports both conventional (nontrunked) radio and trunked-radio system configurations. Figure 5.1 illustrates how a Project 25 system would operate in analog and digital mode [2].

5.2 ARCHITECTURE

Figure 5.2 shows the overall architectural framework, called the "General System Model" in Project 25 terminology. Typically, a framework architecture identifies the functional entities and reference points (e.g. Um). Components and interfaces are the

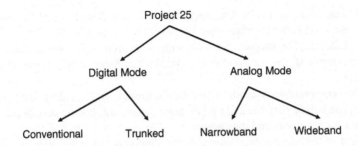

Figure 5.1. Operational modes of P25 radio systems [2].

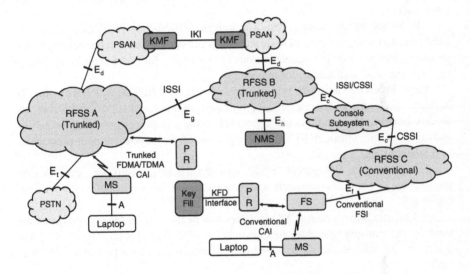

CAI	Common Air Interface	MDP	Mobile Data Peripheral
CSSI	Console Sub-System Interface (Ec)	MDPI	Mobile Data Peripheral Interface
DNI	Data Network Interface (Ed)	MS	Mobile Station
FDMA	Frequency Division Multiple Access	NMI	Network Management Interface (En)
FS	Fixed Station	NMS	Network Management System
FSI	Fixed Station Interface (Ef)	PR	Portable Radio
IKI	Inter Key Management Facility Interface	PSAN	Public Safety Application Network
ISSI	Inter Sub-System Interface (Eg)	PSTN	Public Switched Telephone Network
KDI	Key Fill Device Interface	RFSS	Radio Frequency Sub System
KFD	Key Fill Device	TDMA	Time Division Multiple Access
KMF	Key Management Facility	TII	Telephone Interconnect Interface (Et)

Figure 5.2. Project 25 framework architecture [4].

physical manifestations of the functional entities and reference points, respectively. For example, CAI is the interface specified over the reference point Um.

The model used by Project 25 also includes a few types of subsystems such as the RF Sub-System (RFSS), Conventional Console Sub-System, and Trunked Console Sub-System.

Major components include fixed conventional and trunked base stations, repeaters, subscriber portable and mobile radios, and data specific peripherals as well as network management and data hosts.

Project 25 reference points/interfaces are identified and specified to provide standardized connectivity among the components and systems. Project 25 includes not only the over-the-air interfaces but also a complete suite of interfaces for its wired infrastructure.

The RFSS, which includes one or more base stations (conventional and trunked fixed stations), provides control logic and call processing to support the interfaces and protocols that are used to connect various components. An RFSS may have a single site (i.e. one base station) or multiple sites.

Large Project 25 networks typically include multiple RFSSs, which are primarily building blocks for wide area critical communication networks. Project 25 does not specify how the base stations are connected to the equipment providing switching and control logic within an RFSS. This aspect is left to vendors for their proprietary implementations.

Fixed Network Equipment (FNE), aka the trunking controller, receives user requests to access traffic channels and assigns an available channel to the requesting user.

A Console Sub-System (CSS) includes one or more consoles (devices) that are used by the console operators (dispatchers or supervisors) who interact with the agents in the field. It should be noted that a console may be a part of an RFSS as well.

Fixed Stations (FSs) are base station equipment that include radios and antenna towers with high power transmitters. Similarly, repeaters are high powered radios with antenna towers used to increase the range of radio transmission by merely retransmitting the signals received.

Devices that are used chiefly by the agents in the field that a Project 25 system supports are mobile subscriber radios and data specific peripheral devices. Subscriber radios include hand-held portable radios or mobile radios that are typically attached to vehicles, are powered by the battery of the vehicle, and use a taller antenna attached to the chassis. Data specific peripherals include commercial laptops, aka Mobile Data Peripherals (MDP).

A Key Management Facility (KMF) is essentially a server that facilitates secure encryption key management and distribution. Typically, a KMF can generate, load, and delete encryption keys on-demand.

A Key Fill Device (KFD) is an in-field radio encryption tool providing an error-free and fast entry of the keys into mobile subscriber devices. It typically works with the KMF for updates in the field or remotely via Over-the-Air Rekeying (OTAR).

Public Safety Application Network (PSAN) refers to an agency-specific group of connected application computers, data hosts, providing various services (such as dispatch, answering) within that agency.

5.3 INTERFACES

A Project 25 digital radio system is composed of several open, standardized interfaces connecting an RFSS to other components. However, Project 25 leaves it to the vendor to specify and build the interfaces (e.g. trunk controller-base station interface) inside an RFSS. Table 5.2 shows a list of Project 25 standards interfaces.

TABLE 5.2. Project 25 Interfaces

Interface Category	Interface	Description
Air interfaces	Common Air Interface (CAI)	Radio to radio protocol
Wireline interfaces	Inter Sub-System Interface (ISSI)	RFSS to all other system interconnections
	Console Sub-System Interface (CSSI)	Console to RFSS
	Fixed Station Interface (FSI)	Fixed station to RFSS or to console subsystem
	Network Management Interface (NMI)	Network Management System to the RFSS
	Telephone Interconnect Interface (TII)	PSTN to RFSS
Data interfaces	Data Network Interface (DNI)	Computer-aided dispatch to the RFSS
	Mobile Data Peripheral Interface (MDPI)	Radio to data peripheral
Security interfaces	Inter Key Management Facility Interface (IKI)	Between key management facilities
	Key Fill Device Interface (KDI)	Radio to key fill device

TABLE 5.3. Types of Channels Supported on the CAI

Channel Type	Bandwidth	Phase	How Many
FDMA control channel	12.5 KHz	Phase 1 and 2	1
FDMA voice channel	12.5 KHz	Phase 1 and 2	2 in Phase 1
			1 in Phase 2
FDMA data channel	12.5 KHz	Phase 1 and 2	1
TDMA voice channel	6.25 KHz	Phase 2	2

5.3.1 Air Interfaces

The **CAI** specifies the type and content of signals transmitted between compliant radios over an FDMA or TDMA (Phase 2) channel at a maximum of 9.6 Kbps data rate. Table 5.3 provides a list of different channels supported on the CAI.

As shown in Figure 5.2, CAI may be used to connect a subscriber unit over the air to:

- other subscriber units directly,
- other subscriber units indirectly via a repeater,
- other subscriber units indirectly via a fixed station (conventional or trunked), and
- other subscriber units indirectly via a repeater and a conventional fixed station.

Audio on these systems is exclusively digital using the CAI standard, which includes an Improved Multi-Band Excitation (IMBE™) or Advanced Multi-Band Excitation (AMBE™) vocoder to digitize voice, error correction code, encryption, and other signaling information for reliable and secure transmission of user information. Figure 5.3 shows the structure of an audio message, which is composed of a header, several Logical Data Units (LDUs), and a terminator field. Digitized audio bits are grouped into 20 ms audio frames (audio bits plus error correction bits), which are then placed into the LDUs. Each LDU contains nine audio frames.

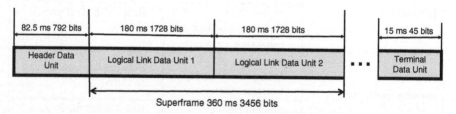

Figure 5.3. CAI audio message fields [2].

Data messages sent over the CAI are divided into packets of varying length. Each packet is then provided an error correction code and a header with packet length information. The Control Channel (CC) and traffic channels are used for data transmissions. The maximum data rate on the CC, the voice FDMA, and data FDMA channels is 9.6 Kbps and on the voice TDMA channel is 12 Kbps.

The CAI is similar for both trunking and conventional Project 25 systems. The following are the significant differences:

- The management of users' access in the trunked version includes a command and response process to communicate with a trunking controller.
- The trunked version in Phase 2 specifies an additional TDMA-based interface with two slots in 12.5 KHz bandwidth (the equivalent of two voice channels).

5.3.2 Wireline Interfaces

The **Inter-RF Subsystem Interface (ISSI)** allows Project 25 systems from different manufacturers to be directly interconnected at the controller level, allowing seamless cross-system intercommunication, and system-to-system roaming for same-band systems. The ISSI is a vital wireline interface for achieving interoperability in multivendor, multi-RF technology (TDMA, FDMA), and multiband Project 25 systems allowing automatic roaming (mobility management) among different networks.

The primary player to make all these possibilities is a comprehensive protocol with its messages (format, content, procedures) running on the ISSI.

With the ISSI, wide area networks with any topology (ring, loop, star, mesh) can be designed and deployed with private and public transmission links.

The **Console Subsystem Interface (CSSI)** is a multichannel digital interface supporting standard protocols to connect dispatch consoles (i.e. CSSs) to the RFSS. The CSSI is an Ethernet-based interface with an RJ45 connector. As shown in Figure 5.2, a CSS can be connected to a fixed station with a Fixed Station Interface (FSI). The CSSI is a subset of the FSI. Furthermore, a CSS may also have a subset of the data host or network interfaces.

The **Network Management Interface (NMI)** connects a network management system to all RFSSs to provide network management support (operations, administration, and maintenance) at all levels (components, subsystems, whole network, and service levels) for the system. More formally, this is an interface between an Operations and Maintenance Center for RFSS (OMC-RF) and a system level Network Management Center (NMC) [5].

Project 25 relies on already-established industry standard protocols in this specific area. Instead, the focus of this interface is to identify features of the existing network management standards, such as Simple Network Management Protocol (SNMP), suitable for Project 25 especially on fault and performance management functional areas.

TABLE 5.4. DFSI Protocol Suite

Control Protocol	Voice Conveyance Protocol	
	RTP	RTCP
	UDP	
	IP	
	Ethernet	

The **Fixed Station Interface (FSI)** connects conventional base stations and repeaters to other components such as an FSSS or a console system. The Conventional Fixed Station Interface (CFSI), a specialization of the FSI, supports both analog and digital connections.

- Analog Fixed Station Interface (AFSI)—used to connect the analog base station to an FSSS or a console via two and four-wire circuits. The AFSI supports full or half duplex or simplex communications as well as group calls.
- Digital Fixed Station Interface (DFSI)—uses the Ethernet with the RJ45 connector, the Internet Protocol (IP), the User Datagram Protocol (UDP), and the Real-time Transport Protocol (RTP) or Real-time Transport Control Protocol (RTCP) shown in Table 5.4. The control and signaling information are carried by the UDP, and the voice is transported by the RTP or RTCP messages. The DFSI supports all conventional Project 25 features, including group and individual calls.

The **Telephone Interconnect Interface (TII)** supports voice-specific communication between an RFSS and the telephone network, more formally the Public Switched Telephone Network (PSTN). The TII supports both analog and digital interfaces. The Integrated Services Digital Data Network (ISDN) telephone interfaces and optionally other interfaces are also supported. With the ubiquitous availability of cellular phones and the Internet, there is no real need for this voice-only interface anymore.

5.3.3 Data Interfaces

The **Data Network Interface (DNI)** supports data-specific communication between an RFSS and a data center or a data network, formally called Public Safety Application Network (PSAN). Typically, a device called data gateway in the RFSS provides this functionality. DNI supports a number of protocols, ranging from a Project 25 specific protocol to the IP protocol suite as well as some legacy protocols such as X.25.

The **Mobile Data Peripheral Interface (MDPI)** supports the link between a subscriber mobile or portable radio unit and a data-specific device such as a laptop.

This is also called the Subscriber Data Peripheral Interface (SDPI). Some open interface protocols are allowed with a transparency requirement, which ensures that the supported protocols are carried over the protocols supported by the interfaces on the RFSS.

5.3.4 Security Interfaces

The **Inter Key Management Facility Interface (IKI)** is used to connect two Key Management Facilities (KMF).

The **Key Fill Device Interface (KDI)** is used between a subscriber mobile or portable radio unit and a key fill device.

5.4 SERVICES

The services available in Project 25 based critical communications systems heavily depend on whether the system is conventional or trunked.

A **conventional Project 25** system offers voice services, supplementary services, added data service, and configuration management over the FDMA CAI and wireline interfaces. On "direct" configuration (subscriber unit to subscriber unit) of the conventional system, voice services include individual, group, and "all" types of calls, "Talking Party ID," and more. Supplementary services include emergency alarm, call alert, status, or message, and more. On "repeated" configuration (subscriber unit to subscriber unit via one or more repeaters in the middle) of the conventional system, voice and supplementary services are similar, except that there may be multiple channels over multiple sites providing voting, multicast, or simulcast features as well. A repeated configuration may also include a wireline dispatch (console), which may also participate in calls, possibly originating and receiving them.

All the conventional configurations mentioned above support some encryption service as well. For example, encryption keys, for transmission, may be selected for each conversation, each channel, or each talk group. Encryption keys may also be preselected on reception, or use an already stored key in the device. An external "key fill device" may also be used to download encryption keys to the subscriber units.

Similarly, added data services on conventional Project 25 systems can be discussed under the three configurations mentioned above. On direct configurations, a data signaling protocol and a data bearer service (transporting IPv4 datagrams) on the FDMA CAI are supported by the applications in the subscriber unit. On repeated configurations, data signals are simply repeated by the fixed stations. However, on configurations involving wireline dispatches and data gateways, data may be exchanged between a subscriber unit and an external data host connected to a data gateway in the RFSS.

All the conventional data configurations discussed above support encryption services in subscriber units and data gateways. For example, packet payloads are encrypted on the air and decrypted before they are routed to the wireline network. Again, an external "key fill device" may also be used to download encryption keys to the subscriber units and data gateways in the RFSS.

A **trunked Project 25** system supports, over the air on the CAI, integrated voice and data services including encryption over end-to-end voice and data communications. Some examples of the supplementary services include priority call, pre-emptive priority call, call alert, emergency alarm, and short message. To facilitate seamless communications while the users roam from one RFSS to another, mobility and registration related services are also available. Upon requests via the control channel, appropriate FDMA data channels are allocated for data transactions, which may include location or OTAR related data services.

The services provided over the ISSI include intersystem group call, broadcast call, intersystem unit-to-unit call, emergency group call, and emergency alarm.

The services provided over the CSSI mainly support the features available on the ISSI. Monitoring simultaneous talk groups and a console taken over by another console are two examples of such services. Also, the CSSI supports several console-specific services such as console-initiated group call, radio-initiated group call, emergency group call, console priority, and support for multiple talk groups [5].

5.5 OPERATIONS

Project 25 facilitates several system configurations supporting the following operations:

- Direct mode
- Repeated
- Single site
- Multisite
- Voting
- Multicast
- Simulcast

This large set of configuration choices provides geographical coverage flexibility to select an option that matches the characteristics of the geographical area. For example, simulcast operation is more appropriate in urban areas with a large population, whereas a Project 25 system with high-power operation can cover large rural areas with a small population with fewer sites.

Project 25 allows conventional operations, which are more suitable for low-density applications in rural areas, and trunked operations (in Phase 2), which are more suitable for high-density applications in urban areas.

As discussed in Chapter 5, in a conventional system, fixed RF channels are used directly by the users, without a control channel. The user selects an RF channel and begins talking directly and immediately, since there is no call setup delay.

In contrast, in a trunked system, a pool of RF channels are shared by the users; a trunk controller, with the help of a control channel, facilitates and manages the sharing and automatic allocation of the RF channels to the users. However, trunked systems may not be suitable for some scenarios where multiple agencies get involved in responding to a disaster situation. In these cases, conventional systems, perhaps with portable repeaters, may provide better, quicker communication among radios in the field.

5.6 SECURITY

An essential component of security in any communication system is the encrypted transmission of signaling and user voice and data traffic. Project 25 is no exception. Digital encryption is an important, but optional, feature specified in Project 25 standards. Users can select this option when it is deemed necessary to have the transmission encrypted.

The CAI supports any of the following four types of encryption algorithms available in the USA:

- Type 1 for classified US material (national security)
- Type 2 for general US federal interagency security
- Type 3 for interagency security among federal, state, and local agencies; both Data Encryption Standard (DES) and Advanced Encryption Standard (AES) are supported in this category
- Type 4 for proprietary solutions systems

A primary element of an encryption algorithm is a "private" key that both communicating entities must use in encrypting and decrypting the information being communicated. Management of these keys, which includes secure creation and dissemination, is typically handled by a Key Management Facility (KMF). The Project 25 standard also includes a function called OTAR, which allows the transfer of encryption keys via radio. This way, encryption keys can be managed and changed remotely. OTAR signaling is sent via packet data units over the CAI.

5.7 RF SPECTRUM

Project 25 supports multiple frequency bands. Depending on the scale and needs, an organization may prefer to use a single frequency band or multiple frequency bands.

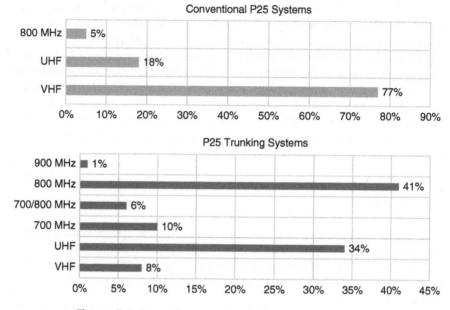

Figure 5.4. Use of frequency bands by Project 25 systems [6].

The frequency bands in which Project 25 radio systems are available are Very High Frequency (VHF) (136–174 MHz) and Ultra High Frequency (UHF) (403–512 MHz, 806–870 MHz). Also, Project 25 Phase 1 technology has been adopted by the Federal Communications Commission (FCC) in the USA as the digital interoperability standard for the new 700 MHz (746–806 MHz) digital public safety band.

As illustrated in Figure 5.4, VHF is more popular among the conventional Project 25 deployments, while the trunked Project 25 system deployments use UHF and 800 MHz frequencies more often [6].

5.8 STANDARDIZATION

Project 25 is a product of collaboration among several organizations, which include the APCO, the NASTD, selected federal agencies, and the NCS in the USA. The specifications are standardized under the TIA.

While APCO is the sole developer and formulator of the standard, TIA provides technical assistance and documentation for the standard. TIA is a national voluntary industry organization of manufacturers and suppliers of telecommunications equipment and services. TIA has substantial experience in the standardization of telecommunications. The formal Project 25 standards development in TIA takes

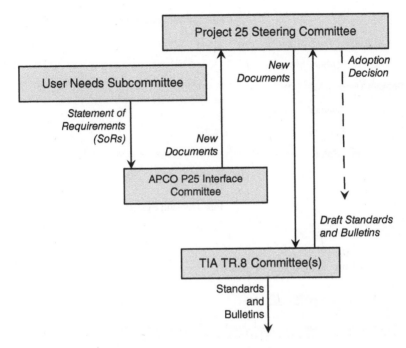

Figure 5.5. Project 25 standardization process.

place in TIA's Mobile and Personal Private Radio Standards Committee (TIA TR-8). The official designation of the Project 25 suite of standards is called the TIA-102 [3].

As depicted in Figure 5.5, Project 25 is directed by a steering committee, which includes experts from various public safety agencies.

The Project 25 Statement of Requirements (Project 25 SoR) is the basis for the APCO-driven standardization process, as well as for the processes used in procuring Project 25 based products from the manufacturers. All activities of the process must be approved by the steering committee and the TIA based on a Memorandum of Understanding (MoU) signed in 1993. The Project 25 Steering Committee specifies the suite of documents that compose the Project 25 Standard.

More than 125 documents support the Project 25 suite of standards (around 75 standard documents and more than 25 TIA Telecommunication Systems Bulletins [TSBs], which are informative documents addressing the background, operational and functional aspects, and guidelines for engineers and users). See Appendix A for a complete list of these documents.

There are additional TIA documents, which describe the suite of tests to demonstrate compliance (conformance, performance, and interoperability) for each core definition document (note that in the USA, there is a formal program called the

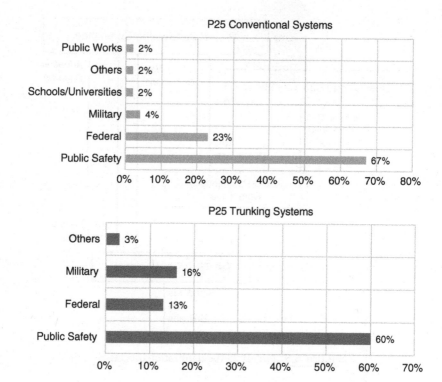

Figure 5.6. User category breakdown of Project 25 systems [6].

Compliance Assessment Program [CAP], which coordinates the process of testing and getting approval from the Department of Homeland Security) [5].

Project 25 continues to evolve. Project 25 is a suite of living documents, which are continuously updated, revised, and published to incorporate new user requirements as well as advances in technology. The ongoing work in APCO, TIA, and other stakeholders has been centered on issues related to the interoperability between Project 25 and Long Term Evolution (LTE) based public safety networks.

5.9 DEPLOYMENT

Project 25 has been adopted in 83 countries around the world (as shown in Figure 5.7) and more than 40 companies are building Project 25 compliant equipment, as shown in Table 5.5.

According to a report by the Project 25 Technology Interest Group (PTIG) [7], an industry consortium established to promote Project 25, as of July 2016, there is a total of 2141 Project 25 systems worldwide (1299 conventional and 842 trunked) [6].

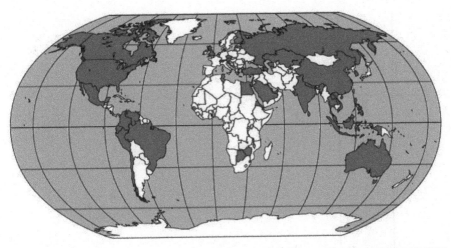

Australia	Canada	El Salvador	Kazakhstan	Philippines	Trinidad
Austria	Chile	Eritrea	Kuwait	Russia	Tunisia
Azerbaijan	China	Finland	Latvia	Saudi Arabia	Turkey
Bahrain	Colombia	India	Laos	Singapore	United Kingdom
Bermuda	Costa Rica	Indonesia	Malaysia	Slovenia	USA
Botswana	Czech Republic	Hong Kong	Mexico	South Korea	United Arab Emirates
Brazil	Ecuador	Jamaica	Nepal	Sri Lanka	Venezuela
Brunei	Egypt		Peru	Switzerland	Vietnam
				Thailand	Zimbabwe

Figure 5.7. Countries with Project 25 based systems [1].

As shown in Figure 5.6, over 60% of the deployed Project 25 systems are used by public safety agencies. The use of these systems by federal agencies as well as the military is also significant.

Project 25 has been developed specifically with public safety agencies in mind. Therefore, its design and features are optimized for this purpose. It is relatively more straightforward than TETRA. Project 25 based public safety networks typically require less number of base stations. Therefore, the overall cost of the network is comparatively less. However, due to the economy of scale, the devices are more expansive. The other cost factors for Project 25 are similar to the ones for TETRA.

5.10 FUTURE

In general, Project 25 is expected to continue to change by incorporating new user needs and new regulatory requirements as well as technological advances. However,

TABLE 5.5. Vendors Offering Project 25 Products and Services [7]

Organizations	Fixed Stations and Repeaters	Mobiles and Portables	Consoles	Networks	Software	Test Equipment	Systems Integration	Consultant Services
Aecom						x	X	x
Aeroflex						x		
Airwave Solutions							X	x
Anritsu						x		
Avtec			x					
Cassidian	x			x			X	
Catalyst			x					
Cisco Systems				x	X			
Cobham Avionics		x		x	X			
Codan	x			x			X	
Radio-Daniels								
Cynergize							X	x
Datron		x						
DVSI					X			
EF Johnson	x	x	x	x	X			
Etherstack				x	X	x		
Federal Engineering					X		X	x
Genesis Group					X		X	
Harris	x	x	x	x			X	
Icom America	x	x		x				
Kenwood USA	x	x		x				

Midland Radio	x					
Moducom	x	x	x		x	
Motorola Solutions	x	x	x		x	x
Pantel International		x	x			
Power Trunk	x	x	x		x	x x
Radio Comm Solutions						x x
Raytheon	x	x	x		x	x
Relm Wireless	x			x		x
Simoco	x	x				x
Spectra Engineering	x					x
Tait Comm.	x	x	x		x	
Technisonic	x					
Telex Radio Dispatch		x	x		x	
Thales		x				
Vertex Standard		x				
Wireless Pacific	x					
Zetron		x	x		x	

there is no plan to evolve Project 25 into a broadband data transport capable system. However, there is a strong desire for it to be a part of next generation LTE-based broadband public safety networks. Therefore, there are some activities to add interoperability features and participate heavily in the development of critical communications features in the LTE standardization process [8].

For Project 25, there was a plan to go to Phase 3 to address high-speed broadband data needs. According to this plan, ETSI and TIA would collaborate under a project known as MESA, short for Mobility for Emergency and Safety Applications [9]. However, due to the latest broadband public safety development in the USA to focus on LTE-based public safety networks and similar activities in Europe, Project MESA was discontinued.

The current focus of the primary stakeholders of Project 25 is on how Project 25 features can be incorporated into LTE-based public safety networks, and thus, how the current Project 25 users can interact with the users with LTE-based terminal devices. There is an ongoing effort to develop a standard for Project 25 Push-To-Talk (PTT) over LTE (Project 25 PTToLTE). The Project 25 community would like to leverage Project 25 features and interfaces, especially the ISSI, as the foundation and push this through to international standards development organizations such as 3GPP. The following are the goals of this effort:

- Interoperable PTT solution between Project 25 and LTE
- Interoperable Project 25 PTT services over LTE
- Leverage the Project 25 ISSI protocol, which is SIP-based, for session management and Real-Time Transport Protocol (RTP) to transport Project 25 voice frames.
- All-IP connectivity, IPv4 based, compatible with IPv6

5.11 SUMMARY AND CONCLUSIONS

The chapter provided a high-level discussion on Project 25, a joint effort among several government agencies, user groups, and industry associations in the USA. Technical specifications, network architecture, interfaces and protocols, and services are discussed to a certain degree without going into the nitty gritty of the standards and other official documents. The standardization process and its future are also discussed.

Project 25 is widely deployed in the world and multiple vendors offer multiple Project 25 products.

Project 25 is expected to continue to evolve to respond to new requirements. However, it is also expected that broadband-based capabilities will not be a part of future Project 25 specifications due to a consensus that such capabilities will be handled by LTE-based solutions.

REFERENCES

1. "Project 25." [Online]. Wikipedia. Available: https://en.wikipedia.org/wiki/Project_25. [Accessed: Dec. 12, 2016].
2. Codan Radio Communications, *P25 Radio Systems Training Guide, Revision 4.0.0.* Codan Ltd., 2013.
3. Telecommunications Industry Association, "TIA-102_Series_Documents." [Online]. Available: https://archive.org/details/TIA-102_Series_Documents. [Accessed: Sep. 15, 2018].
4. PTIG—The Project 25 Technology Interest Group, "Project 25 foundations panel discussion for 2016," presented at Int. Wireless Commun. Expo., Las Vegas, Mar. 21, 2016.
5. PTIG—The Project 25 Technology Interest Group, "Project 25 foundations panel discussion for 2015," presented at Int. Wireless Commun. Expo., Las Vegas, Mar. 16, 2015.
6. "PTIG publishes new P25 systems list with P25 conventional for June 2016." [Online]. Available: http://www.project25.org/index.php/news-events/338-ptig-publishes-new-p25-systems-list-with-p25-conventional-for-june-2016. [Accessed: Jan. 9, 2017].
7. "Project 25 Technology Interest Group." [Online]. Available: http://www.project25.org/index.php. [Accessed: Jan. 10, 2017].
8. D. Tuite, "Can public-safety radio's P25 survive LTE?" [Online]. Electronic Design, Jul. 17, 2012. Available: https://www.electronicdesign.com/analog/can-public-safety-radio-s-p25-survive-lte.
9. "Spectrum management for the 21st century: The President's spectrum policy initiative," National Telecommunications Information Agency, Mar. 2008.

6

TERRESTRIAL TRUNKED RADIO (TETRA)

This chapter presents Terrestrial Trunked Radio (TETRA), another narrowband Land Mobile Radio (LMR) technology that provides voice and low-speed data communication capabilities for critical communication systems. The chapter discusses TETRA's technical specifications, network architecture, interfaces and protocols, and services. The standardization process and its future are also discussed. Appendix B provides a comprehensive list of related standard documents.

6.1 INTRODUCTION

TETRA is a critical communication standard developed by European Telecommunications Standards Institute (ETSI). It became widely used in many countries around the world [1–3]. Before we go further on TETRA, let's remind the reader that there is another digital LMR standard called TETRAPOL [4], which is not related to TETRA, other than the use of the acronym "TETRA," as discussed briefly in Chapters 1 and 4.

Fundamentals of Public Safety Networks and Critical Communications Systems: Technologies, Deployment, and Management, First Edition. Mehmet Ulema.

© 2019 by The Institute of Electrical and Electronics Engineers, Inc. Published 2019 by John Wiley & Sons, Inc.

The TETRA project began in the 1980s with an objective to develop a standard for a commercial wireless mobile network. Because of this objective, initial specifications included many features like traffic maximization. While ETSI was spending many years in developing this comprehensive system, another ETSI standard, Global System for Mobile Communications (GSM), became popular and ubiquitous, dampening the TETRA effort. This setback caused significant hardship for the companies invested in TETRA development, and they identified the public safety market as a way to sell their products. With the support of several European public safety agencies, TETRA became the most popular solution for public safety in Europe and around the world.

Note that when we say "TETRA standard," we refer to a set of related standards developed by ETSI. See Appendix B for a comprehensive list of active TETRA standards. Also, note that TETRA standards have been developed in phases over time. ETSI uses the word "release" to refer to these phases—TETRA Release 1 and TETRA Release 2.

TETRA allocates channels to users on demand in both voice and data modes. Additionally, a few national and multinational networks are available and national and international roaming can be supported. The systems makes use of the available frequency allocations using Time Division Multiple Access (TDMA) technology with four user channels on one radio carrier with 25 kHz spacing between carriers.

TETRA Enhanced Data Service (TEDS), included in TETRA 2, enables more data bandwidth to TETRA data service users. The standard allows up to 691 Kbps, but limitations in spectrum availability typically give users a net throughput of around 100 Kbps. TETRA 2 also includes additional features such as enhanced cell radius for air-to-ground communications and Location Information Protocol (LIP) for advanced automatic location services.

Some of the critical features of TETRA are listed in Table 6.1. The following are some additional significant features:

• Air interface plus end-to-end encryption
• Mutual authentication
• Integrated voice
• Multislot packet data with pre-emption
• Unlimited number of users supported

6.2 ARCHITECTURE

As shown in Figure 6.1, the TETRA architecture features a number of standardized interfaces and allows some vendor proprietary interfaces, as discussed in Section 6.3 in more detail.

TABLE 6.1. Main Parameters of TETRA [5]

Parameter	Value
Frequency bands	Several bands 380–460 MHz
Carrier spacing	25 kHz
Duplex method	FDD
Modulation	QPSK
Carrier data rate	36 Kbit/s
Access method	4-slot TDMA per 25 kHz channel
User data rate	7.2 Kbit/s per time slot (unprotected)
Voice coder rate	ACELP (4.56 Kbit/s net, 7.2 Kbit/s gross)
Power	Up to 40 watt base stationUp to 1.8 watt portable,3–10 watt mobile
Texting	Up to 1000 characters

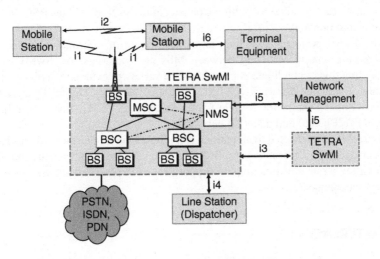

i1: Air Interface

i2: Direct Mode Interface

i3: Intersystem Interface (ISI)

i4: Line Station (Dispatcher) Interface

i5: Network Management Interface

i6: Terminal Equipment Interface

PSTN: Public Switched Telephone Network

MSC: Mobile Switching Center

BS: Base Station

BSC: Base Station Controller

NMS: Network Management System

SwMI: Switching and Management Infrastructure

PDN: Public Data Network

Figure 6.1. TETRA network architecture.

A typical TETRA network architecture consists of one or more Switching and Management Infrastructures (SwMIs) connected by trunks, aka a backhaul transmission network. SwMIs, also called TETRA nodes in some literature, support Mobile Stations (MSs) via air (radio) interfaces, and Line Stations (LSs), commonly known as "dispatchers." A network management center is connected to every SwMI in the network to provide operations, administration, and maintenance functions. Terminal equipment, mainly laptop and handheld computers, can be connected to mobile stations via directly wired interfaces as well.

An SwMI consists of Base Stations (BSs) supporting radio transceiver towers, Base Station Controllers (BSCs), Mobile Switching Centers (MSCs), and a Network Management System (NMS). The BS resends information from an MS to the requested receiver. Several BSs are connected to a BSC; in turn, some BSCs are connected to an MSC. These hierarchical connections and the number of entities connected vary significantly depending on the size of the network. Also, we should note that some vendors have products that combine these functionalities into a single unit. For example, the so-called Switching Control Node (SCN) combines BSC and MSC functions into one unit.

The SwMI performs switching and transmission of information (voice, data) and signaling among MSs and between MSs and gateways. An SwMI includes gateway functions to facilitate connections to a variety of external telephony (Public Switched Telephone Network [PSTN], Integrated Services Digital Data Network [ISDN], Global System for Mobile Communications [GSM], Voice over Internet Protocol [VoIP], etc.) and data (IP) networks.

The NMS is connected to all entities in an SwMI via IP interfaces to provide network management applications. An SwMI may have some other systems and servers for providing a range of functions to support, for example, roaming, authentication, and key management.

6.3 INTERFACES

As shown in Figure 6.1, the TETRA architecture includes several air interfaces and network interfaces standardized for TETRA applications, as discussed below. TETRA also has some additional interfaces, especially within the SwMI, which are left to the vendors to specify and implement.

6.3.1 Air Interfaces

TETRA specifies two types of air interfaces corresponding to the two fundamental operations involving the base stations and mobile stations:

- **Trunked Mode Operation (TMO):** The air interface in this operation refers to the radio interface between a base station and the mobile stations

TABLE 6.2. Bandwidth on Demand (Kbps)

| | Number of Time Slots | | | |
	1	2	3	4
Protection Level				
No Protection	7.2	14.4	21.6	28.8
Low Protection	4.8	9.6	14.4	19.2
High Protection	2.4	4.8	7.2	9.6

communicating through this base station. The following is a list of primary characteristics of an air interface operating in this mode:

– TDMA structure

– Four user channels

– 25 kHz channel spacing

– 36 Kbps transmission rate

– Π/4 Differential Quadrature Phase Shift Keying (DQPSK) modulation

– 7.2 Kbps speech coding (including error correction)

Effective data rate capacities that can be achieved on the air interface operating in trunk mode depend on the protection level demanded by the application, as shown in Table 6.2. The types of applications that may be using this "bandwidth on demand" future include real-time circuit mode data applications such as video and slow scan television, and those applications requiring increased throughput (or speed) in packet modes such as file transfer and database updates.

Figure 6.2 shows the protocol stack used for trunk mode operations on the air interface. All services use the bottom two layers (physical and the Medium Access Control (MAC) part of the data link layer). Voice services and circuit

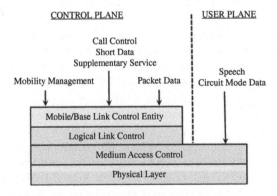

Figure 6.2. Air interface protocol stack for TM operations.

mode data services use only these two common layers. Other services use, via TETRA, defined Application Programming Interfaces (APIs), with their corresponding protocols at Layer 3.

- **Direct Mode Operation (DMO):** The air interface in this operation refers to the radio interface between any two mobile stations communicating directly without any base stations involved. TETRA identifies the following scenarios for direct mode operations:

  o *Basic Scenario*: Any individual (one-to-one) or group calls, as well as broadcast calls, are included in this scenario.

  o *Dual Watch*: While the MS is in the direct mode, the control channel of the trunk mode is monitored. Similarly, while the MS is in the TM mode, DM channels are monitored.

  o *Managed Direct Mode*: This scenario involves a managed (i.e. authorized) operation. Typically, an MS periodically broadcasts the authorization signals.

  o *Repeater*: In this scenario, the MS acts as a repeater that can support two calls simultaneously or a single call on different uplink and downlink frequencies on the same frequency for both uplink and downlink.

  o *Gateway*: In this scenario, the MS interconnects other MSs operating in the DM and TM modes.

  o *Managed Repeater, Gateway, or both Combined*: In this scenario, the MS broadcasts the authorization signals to be used in the repeater and gateway scenarios.

6.3.2 Intersystem Interface

As the name implies, this standardized Intersystem Interface (ISI) is used to interconnect all the SwMIs in a TETRA network. The ISI supports both circuit mode and packet mode information transmissions as well as cross-border roaming. As far as services are concerned, the ISI supports individual calls, group calls, supplementary services, and data services along with mobility management and security features. Some of the mobility management features include migration, deregistration, group attachment/detachment, and group linking/unlinking. TETRA authentication and Over The Air Rekeying (OTAR) for key management are two of the primary security features.

Figure 6.3 shows the ISI protocol stack. The physical layer is based on digital 64 Kbps connections.

The ISI uses Private Signalling System 1 (PSS1) at its higher layers. The PSS1, commonly known as QSIG, is a well-known signaling protocol typically used in private corporate networks mainly among Private Branch Exchanges (PBXs). For the ISI, the PSS1 includes International Organization for Standardization (ISO)/ International Electrotechnical Commission (IEC) 11582, which defines the

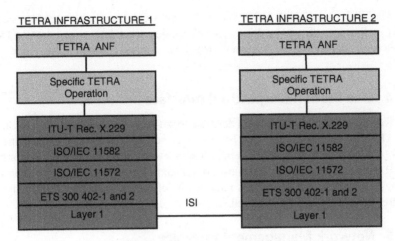

Figure 6.3. ISI protocol stack.

signaling protocol for the control of supplementary services and Additional Network Features (ANFs), with the help of International Telecommunications Union-Telecommunications sector (ITU-T) Rec. X.229 (protocol specification for remote operations), as shown in Figure 6.3. Furthermore, ISI Layer 3 is covered by the ISO/IEC 11572, which defines interexchange signaling procedures and protocol for circuit mode bearer services. At the data link layer, the ISI uses ETS 300 402-1 and 2, which are based on the ISDN.

6.3.3 Terminal Equipment Interface (TEI)

This interface is used to exchange data between a mobile station and terminal equipment. The terminal equipment, also called peripheral in some publications, is an external device such as a laptop computer hosting one or more applications that communicate with their counterparts residing in mobile stations. These applications may be involved in controlling various parameters of mobile stations. To facilitate this, several so-called "access services" are offered on this interface. These are provided via:

- Attention (AT) commands
    o Used to access and control circuit mode data services, short data services, and radio configuration related parameters
- TETRA Network Protocol 1 (TNP1)
    o Used to access parameters related to Circuit Mode Control Entity (CMCE) and radio configurations in mobile stations

Packet data service is accessed directly via the Internet Protocol (IP) on top of the Point-to-Point Protocol (PPP).

The TEI uses ITU-T V.24/V.28 at the physical layer with speeds less than 20 Kbps. Services offered by this interface include packet data, circuit data, and short data services depending on the capabilities of the terminal equipment, as well as voice calls.

6.3.4 Line Station (Dispatcher) Interface

TETRA identifies this interface but does not specify it for standardization. This means that TETRA recognizes the need for an interface between the SwMI and line stations (mainly remote dispatcher systems). However, given the existing vendor base, TETRA also recognizes the difficulties of reaching a consensus for a common standardized interface. Therefore, this interface is left unspecified by TETRA; instead, vendor-provided interfaces are used.

6.3.5 Network Management Interface

TETRA does not specify this interface; instead, it provides a set of guidelines, incorporated into the Designer's Guide, for TETRA network administrators and vendors. The guidelines address internal and external aspects of managing TETRA networks.

Internal network management refers to the management (monitoring and controlling) of a single or multisite TETRA network. An internal network management system is assumed. The interfaces and protocols between this network management system and other components are not specified. The choices, such as Simple Network Management Protocol (SNMP) and Common Management Information Protocol (CMIP), are left to the vendors and operators.

External network management refers to the management of multiple TETRA networks connected by the ISI. One or more external management systems, mainly providing a limited set of overall management functions, such as configuration and fault management, may be involved to facilitate interoperability among various vendors and to manage roaming among multiple TETRA networks.

6.3.6 PSTN/ISDN/PDN

As discussed briefly in the architecture section, an SwMI includes several gateways to interface with external telephony and data networks. Although the interfaces and protocols between a gateway and other internal components are vendor specific, TETRA identifies standardized interfaces between a gateway and its corresponding external network.

A Public Switched Telephone Network (PSTN) gateway provides point to point or multipoint full duplex speech in clear mode with echo cancellation. However, there is no support for supplementary services on this interface. The signaling is maintained, and numbering conversion is provided across the gateway.

An ISDN gateway supports both basic and primary rates with Q.933 Digital Subscriber System no. 1 (DSS1) signaling [6]. The ISDN gateway converts TETRA signaling to DSS1 signaling and vice versa. The gateway supports rate adaptation to 64 kbit/s PCM and a small subset of supplementary services. However, this interface does not support circuit mode data services. Thanks to this interface, a user on the ISDN side can reach a TETRA user via ISDN supplementary services or by using, although not always convenient, two-stage dialing.

A Packet Data Network (PDN) gateway supports an interface, called the IP Interworking (IPI) interface, between the TETRA network and a PDN. General Packet Radio Service (GPRS) interfaces of the GSM cellular system are used on this interface, either directly or with an IP-based network in the middle. GPRS tunneling protocol is used to support roaming where the users' roaming profiles are moved through the ISI between databases.

6.4 SERVICES

Some of the critical services provided by TETRA include individual simplex and duplex calls, group calls, pre-emptive emergency and priority calls, dynamic group number assignment, call authentication, late entry, voice encryption, and packet data services. Low-speed packet data as well as circuit data modes are available, along with some form of encryption. All these services and more offered by TETRA, collectively called Voice + Data (V + D) services, can be discussed under three major categories— basic voice services, supplementary voice services, and data services, as presented in the following sections.

6.4.1 Basic Voice Services

The services in this category include individual calls, group calls, acknowledged group calls, and broadcast call services.

Individual call service refers to one-on-one (point-to-point) basic voice communications between two users. The connection management (i.e. call setup, call disconnect, etc.) is an essential part of this service. Both full duplex and half duplex communications in clear or encrypted modes are supported.

Group call service refers to one-to-many (point-to-multipoint) calls, typically initiated by one user to several other users in the group. The call is established immediately without waiting for an acknowledgment from the other users. This is a "Push to Talk (PTT)" type call allowing a fast, reliable way of sharing information among a group of users.

Acknowledged group call service is similar to the group call service. The difference is that the call may not be established unless a sufficient number of acknowledgments are received by the calling user.

Broadcast call service is also similar to the group call service except that the communication is unidirectional. That means that only the calling user can talk and others can only listen!

6.4.2 Supplementary Services

A supplementary service can modify or supplement a basic service. TETRA standards specify a rather rich set of mission-critical supplementary services, as briefly discussed in the following paragraphs.

Preemptive priority call (also known as an emergency call) supplementary service assigns the highest priority to a call. This means that, when needed, the lowest priority calls can be dropped to accommodate the emergency call. Typically, the calling user initiates this call by using an appropriate function (pulling a switch or pushing a button) on the end user device. Needless to say, the related dispatcher, infrastructure equipment, and other users in the group will be alerted about the priority level of this call accordingly.

Call retention supplementary service is used to protect calls against preemption, enabling users to maintain their calls when the highest priority call is guaranteed.

To ensure this, TETRA uses an internal scheme by assigning a Call Retention Value (CRV) to calls based on specific criteria such as the lifetime of the call, the type of call, and the user. When it becomes necessary, the lowest CRV valued call is dropped.

This service may be initiated either by a calling user or a called user. This may be done through the call setup by the calling user or may be requested by a called user for an incoming call.

Priority call supplementary service is used to assign to each call a priority level, which is used by the infrastructure to allocate appropriate network resources to calls based on their priority level.

TETRA allows 16 different priority levels, which can be indicated either during initial call setup by the calling user or could be determined by the network. Higher priority levels indicate the importance of the call and are typically used or given to the agents on the scene.

Dynamic Group Number Assignment (DGNA) supplementary service is used to create special user groups charged with special assignments typically related to incident-specific communications. For example, different users from different organizations could be part of a unique group handling communications to coordinate and manage an emergency situation.

DGNA service may be applied to all the users with ongoing calls or any users and groups with no active calls.

Ambiance listening supplementary service allows a terminal to act as a remote monitor during a critical situation to gather and disseminate information as an ambient (background) noise. This may be established by the calling user or by the dispatcher, who then can listen to ambient noise as well as the voice call being exchanged. Since this call may be part of a group, extreme caution concerning security must be taken when using this service.

Call authorized by dispatcher supplementary service provides a means to the dispatcher to verify a call request before the call is set up. This service may be useful in managing network congestions by allowing only the necessary calls.

Area selection supplementary service provides flexibility for users to select a geographical area for their outgoing calls. This flexibility is also available for incoming calls, which are accepted only when the calling user is situated in a predefined geographical area.

Late entry supplementary service is an air interface feature that allows latecomers to join a group call.

6.4.3 Data Services

Two different types of data services are offered by TETRA—Short Data Services (SDS) and Packet Data Services (PDS). The SDS uses TDMA time slots on the control channel to send, as the name implies, user-defined or predefined short text messages, which can be either 16, 32, 64, or 2048-bit long. On the other hand, PDS is used to send messages with any length.

Short data service, in turn, may be used for individual (point-to-point) or group (point-to-multipoint) messages. With this service, the message is transmitted immediately, making it highly valuable for low-latency applications such as transmitting status and location information. For example, TETRA Location Information Protocol (LIP) uses this service.

Packet data service may be a connection-oriented or a connectionless packet data service. The connection-oriented packet data service first establishes a virtual (logical) connection between the source and the destination and then delivers the data packets. The connectionless service begins the data transfer without establishing any virtual connection between the source and the destination. These two types of packet delivery are well-known techniques in the field of packet-switching networks. The connection-oriented service is considered more reliable and typically more suitable for point-to-point communications, whereas the connectionless service is less reliable and used for either point-to-point or point-to-multipoint communications. For example, TETRA supports IP versions 4 and 6 by using the connectionless packet data service and ITU-T X.25 protocols by using both connection-oriented and connectionless services.

One or more (up to four) TDMA information channels can be used to support packet data services, resulting in between 4.8 Kbps and 19.2 Kbps data rate.

6.5 OPERATIONS

As discussed previously, TETRA offers V+D services via TMO and DMO. These operations are already discussed in some detail in Section 6.3.1. Furthermore, the operations supporting packet data services in TETRA are discussed in Section 6.4.3.

Like many other narrowband technologies used for critical communications systems, TETRA also supports both conventional and trunked operations to provide suitable and efficient configuration options to the network operators (see Chapter 4 for a brief general discussion on these options). However, keep in mind that TETRA is designed mainly to be a trunked system. After all, the extended version of the acronym TETRA is Terrestrial Trunked Radio! Although it may be possible, it may be somewhat costly and inefficient to use TETRA in a conventional operation mode.

6.6 SECURITY

TETRA standards specify a relatively comprehensive set of features related to the user and network security, as discussed briefly here.

- **Mutual authentication** refers to the fact that the TETRA network and the terminal authenticate each other by using a TETRA Authentication Algorithm called TAA1, fully specified in TETRA standards.
- **Encryption** in TETRA can be used on the air interface between the terminal and the network and between two terminals (end-to-end).
  - o *Air interface encryption* is used to encrypt the radio signals transmitted on the air by using one of the four TETRA Encryption Algorithms (TEAs). Alternatively, a country-specific encryption algorithm may also be used for this purpose.

    TETRA defines the following three security classes: Class 1—no encryption, Class 2—static cipher key encryption, and Class 3—dynamic cipher key encryption (with individual, common, or group cipher keys). Authentication in Classes 1 and 2 is optional but is required in Class 3.
  - o *End-to-end encryption* enables two end-user devices to communicate by using an encryption algorithm. Encrypted signals are then transported by the network. The encryption algorithm used must be in compliance with the Security and Fraud Prevention Group (SFPG) of the TETRA Association. A well-known algorithm called the Advanced Encryption Standard (AES) is freely available, and therefore many suppliers include it in their terminals.
- **Secure enabling and disabling of terminals remotely** is a valuable feature. This way, missing terminals and compromised terminals can be disabled.

6.7 SPECTRUM

Theoretically, TETRA can be used in any frequency band from 100 to 1000 MHz (as per the current standard). However, the actual frequency used in a TETRA-based

TABLE 6.3. TETRA-Specified Spectrum Options

Frequency Band	Duplex Space
350–370 MHz	10 MHz
380–400 MHz	10 MHz
410–430 MHz	10 MHz
450–470 MHz	10 MHz
806–824 and 851–869 MHz	45 MHz
870–876 and 915–921 MHz	45 MHz

network depends on the frequency allocated to that region by the regulatory authorities and the availability of equipment supporting that frequency range. TETRA standard specifies in detail the frequency bands, as shown in Table 6.3. Others are not specified in detail.

6.8 STANDARDIZATION

TETRA standards development began in the 1980s by a group of radio manufacturers in Europe under ETSI. To keep up with user requirements, the TETRA work in ETSI went through an evolutionary path. There are two releases of the TETRA standard so far:

- TETRA Release 1—the initial set of specifications
- TETRA Release 2—this 2005 release introduced some new features:
  o Integrated TEDS for high-speed data services
  o Mixed Excitation Liner Predictive, enhanced (MELPe) voice codec
  o Adaptive Multiple Rate (AMR) voice codec
  o MO range extension
  o Enhanced cell radius for air–ground–air
  o Location Information Protocol (LIP) for advanced automatic location services

TETRA in ETSI is expected to continue to evolve beyond Release 1 and Release 2 to provide enhancements only. In other words, ETSI has no plans to develop new technology in this area. The TETRA community has been active in moving toward Long Term Evolution (LTE) based public safety networks as well. Some projects are underway to achieve seamless interoperability between TETRA and LTE-based public safety networks [7].

6.9 DEPLOYMENT

TETRA is in use all around the world, serving more than 2 million public safety officials as well as commercial users. Annual TETRA equipment market value is around 1 billion Euros. It has been adopted for national public safety systems in most countries in the European Union, including Austria, Belgium, Denmark, Estonia, Finland, Germany, Greece, Hungary, Italy, Ireland, Lithuania, Netherlands, Norway, Portugal, Sweden, England, and more.

The TETRA and Critical Communications Association (TCCA) estimates that more than 250 TETRA networks in more than 120 countries are currently deployed by government agencies for public safety, military, and other public services, as well as by other sectors such as utilities and transportation, around the globe.

TETRA is a mature technology. Many companies are involved in this standard. Therefore, there is a wide range of TETRA products available at the current time. These include infrastructure products, which range from small systems to scalable systems that can provide full coverage for a country. There are also many suppliers making terminals. Current terminals are now smaller, with built-in GPS and color screens, and longer battery life.

Major manufacturers have been involved in the standardized and made interoperable radio platforms, so it is easy for a customer to select the best supplier (the European Aeronautic Defense and Space (EADS) Company and Motorola have been the market leaders). Hundreds of successful installations provide confidence about the stability of the TETRA solution. There are also some Asia Pacific manufacturers who develop TETRA products (Table 6.4).

6.9.1 Cost Factors Impacting TETRA Wireless Systems

TETRA was initially designed to cover the commercial wireless mobile market. Consequently, it is a somewhat comprehensive and sophisticated technology. Therefore, the complexity is reflected in development costs; thus, the final product cost.

Another factor that impacts the cost is the trunking approach, which requires high capacity and performance switch nodes in the core network to run the full protocol. It requires high capacity transmission lines connecting base stations and the nodes to avoid excessive signaling delays.

Compared to analog systems, TETRA requires twice the number (or more) of sites to cover the same area covered by the existing analog technology. This means that higher initial investment, as well as recurring periodic costs, is required to handle maintenance, site rent, frequency licenses, and backbone links.

Also, a linear modulation used in TETRA requires more expensive hardware and implies higher energy consumption and dissipation.

Despite all these seemingly negative factors, TETRA is often a cost-effective choice for medium-to-high capacity trunked networks with high traffic volumes and small coverage areas.

TABLE 6.4. Vendors Offering TETRA Products and Services [8]

Organization	Core Products	Peripheral/ Components	Application Provider	Complete Solutions	Integrator	Testing	Consultants
Artevea	X				X		
Abiom Group	X	X				X	
Aeroflex						X	
Agurre	X			X			
Airbus Defence & Space	X	X		X	X		
Air-Lynx	X						
Airwave Solutions Ltd.					X		
Alcatel-Lucent	X				X		
Anritsu Company					X	X	
APD Communications			X		X		
APSI	X	X					X
Arico Technologies							X
Arpeggio Ltd.							
Atos AG					X		
Briscoe Technologies							X
CLEARTONE Telecoms Ltd.	X						
CML Microcircuits		X					
Combilent A/S		X					
DAMM	X						
Eastern Comm. Co. Ltd.	X						
Entropia Digital		X			X		

TABLE 6.4. (Continued)

Organization	Core Products	Peripheral/ Components	Application Provider	Complete Solutions	Integrator	Testing	Consultants
ETELM	X						
Etherstack Limited			X				
Eurofunk Kappacher			X	X			
FREQUENTIS AG	X						X
Funk-Electronic Piciorgros	X						
Huawei Technologies	X	X					
Hytera Mobilfunk GmbH	X		X				
Insta DefSec			X				
Kenwood Electronics	X						
KPN Critical Communications			X		X		
Mentura Group Oy			X				X
Motorola Solutions	X	X	X				
National Instruments						X	
NEC Corporation	X						
Nokia Networks	X	X					
Orbion Consulting Oy							X
P3 communications							X
Panorama Antennas Ltd.		X					
PMR-R&D GmbH	X						
Portalify Ltd.			X				
Prescom			X				
RCS Telecommunications					X		

Company	1	2	3	4	5	6	7
Rheinmetall Defence Electron					X		
Rohde & Schwarz	X	X					X
Rohill Technologies B.V.	X						X
Rolta India Ltd.	X						
SAAB AB	X		X				
SatCom IRL	X		X				
Sectra Communications		X				X	
SELECTRIC Nachrichten-Syst.	X						
Selex ES S.p.A.							X
Sepura plc							X
Siemens Schweiz AG							X
Simoco EMEA Ltd.			X		X		X
SITA SC			X				
ST Electronics Pte Ltd.	X		X				X
Swissphone Wireless AG			X		X		
Teltronic S.A.U.				X	X		X
Testing Technologies				X	X		
TETRAsim (Beaconsim Oy)		X			X		
THALES COMM & SECURITY			X				X
Unimo Technology Co., Ltd.	X						X
Warren Systems	X	X					
XPro Oy	X	X					
Zetron Inc					X		
ZTE							X

Also, TETRA is a mature time-tested standard; many products by many different vendors, applications developers, and expertise are readily available. It is a highly competitive market, which reflects in the cost of products as well as flexibility in the acquisition process. Compared to Project 25 based devices and equipment, TETRA-based devices and equipment seem to be more cost-effective.

6.10 FUTURE

The TETRA community believes that basic Private Mobile Radio (PMR) requirements will remain the same. Therefore, they can be applied to commercially available LTE technology. This implies that TETRA applications and functionalities will be running on top of the LTE telecommunications layer. In other words, LTE technology has been selected as the basis for the next step of TETRA [4, 9–12]. Obviously, new standardization effort is required to provide specifications for end-to-end functionalities and interoperability. LTE technology will enable broadband data applications. The vision also includes support for migration from existing narrowband PMR technologies such as TETRA and Project 25. The TETRA community is concerned that the standardization of the next step is somewhat challenging since it requires a global standard migrating many existing PMR technologies, requiring the collaboration of many standard bodies and stakeholders. Some other concerns include frequency allocation, LTE communities' seriousness about incorporating mission-critical voice services, new business models, and financial aspects.

TETRA community believes that the evolution from narrowband TETRA to LTE-based Broadband PMR is a long process, and narrowband TETRA is expected to remain in operational use until 2025–2030 [11, 13]. In addition to the technical standards development in ETSI, there is an industry association called the TCCA, which is charged to "develop, promote and protect" TETRA technology [8]. Mobile broadband data is their current focus; their aim is to create a single common standard that can be used for mission-critical mobile broadband solutions. TCCA's Critical Communications Broadband Group (CCBG) is tasked with driving the development of common, mobile broadband standards and solutions for critical mobile broadband users worldwide based upon LTE. They have been collaborating with user groups, ETSI, and 3GPP [13, 14].

The National Public Safety Telecommunications Council (NPSTC) and other organizations in the USA recognized this and decided in 2009 on LTE as their platform for their future national public safety network. The USA has reserved a spectrum band in the 700 MHz area for an LTE-based public safety network and, in early 2012, committed $7 billion in funding.

In addition to the further enhancement and "tune-up" work going on in these areas, there is now a clear global consensus that LTE is the technology for next-generation broadband public safety networks.

In Europe, TETRA with its TEDS extension already supports relatively higher data rates up to 100 Kbps, but it is recognized that it is not enough and therefore a new technology is needed to add real mobile broadband capabilities. The TCCA established an objective of "driving the development of mobile broadband solutions, possibly LTE based, for the users of mission critical and business critical mobile communications."

Both the TCCA and NPSTC have publicly stated that they would work with 3GPP to include the functionality necessary within the LTE standard to meet public safety needs. With NPSTC, TCCA, ETSI Technical Committee TETRA, and other organizations backing LTE, there is now a clear global consensus about the use of LTE technology to build next generation broadband public safety networks.

6.11 SUMMARY AND CONCLUSIONS

This chapter provided a relatively extensive coverage of TETRA technology, related standards, architecture, interfaces and protocols, and services.

TETRA developed by ETSI became widely used in Europe and many countries around the world. Additionally, a few national and multinational networks are available and national and international roaming can be supported.

TETRA allocates channels to users on demand in both voice and data modes. TEDS, included in TETRA 2, enables more data bandwidth for TETRA data service users. TETRA 2 also includes additional features such as enhanced cell radius for air-to-ground communications and advanced automatic location services.

TETRA network architecture consists of one or more TETRA nodes, which support mobile stations and dispatchers. A network management center is connected to every TETRA node in the network to provide operations, administration, and maintenance functions.

TETRA is expected to continue to evolve to provide enhancements only. Several projects are underway to achieve seamless interoperability between TETRA and LTE-based public safety networks.

REFERENCES

1. "Terrestrial Trunked Radio (Tetra) in The Network Encyclopedia," 2013. [Online]. Available: http://www.thenetworkencyclopedia.com/entry/terrestrial-trunked-radio-tetra/. [Accessed: Jul. 19, 2016].
2. J. Dunlop, D. Girma, and J. Irvine, *Digital Mobile Communications and the TETRA System.* John Wiley & Sons, 2013.
3. P. Stavroulakis, *TErrestrial Trunked RAdio—TETRA.* Springer Science & Business Media, 2007.
4. "Home | Tetrapol Forum." [Online]. Available: http://www.tetrapol.com/. [Accessed: Jul. 24, 2016].

5. "Global Mass Transit: TETRA technology: Open standard for communication." [Online]. Available: http://www.globalmasstransit.net/archive.php?id=16912. [Accessed: Mar. 4, 2017].

6. "Q.933: ISDN Digital Subscriber Signalling System No. 1 (DSS1)—Signalling specifications for frame mode switched and permanent virtual connection control and status monitoring." [Online]. Available: https://www.itu.int/rec/T-REC-Q.933-200302-I/en. [Accessed: Apr. 18, 2017].

7. B. Mattsson, "TETRA News—TETRA evolution for future needs," TETRA Applications, Mar. 17, 2014. [Online]. Available: http://www.tetra-applications.com/27885/news/tetra-evolution-for-future-needs. [Accessed: Jul. 24, 2016].

8. "TETRA + Critical Communications Association." [Online]. Available: http://www.tandcca.com/. [Accessed: Jul. 24, 2016].

9. "TCCA liaison to 3GPP SA on group communications and proximity services," TCCA, 3GPP SP-120456, Jul. 2012.

10. TETRA and Critical Communications Association, "TCCA signs Market Representation Partner agreement with 3GPP," Jun. 19, 2012. [Online]. Available: https://tcca.info/tcca-signs-market-representation-partner-agreement-with-3gpp/. [Accessed: Sep. 15, 2018].

11. TETRA and Critical Communications Association, "TETRA and LTE working together," Version 1.1, White Paper, Jun. 2014.

12. Motorola Solutions, "The future is now: Public safety LTE communications," White Paper, Aug. 2012.

13. TETRA and Critical Communications Association, "Mission critical mobile broadband: Practical standardization & roadmap considerations," White Paper, Feb. 2013.

14. "3GPP." [Online]. Available: http://www.3gpp.org/. [Accessed: Jul. 24, 2016].

7

DIGITAL MOBILE RADIO (DMR)

This chapter provides an overview of Digital Mobile Radio (DMR) technology, a low-cost replacement of analog systems. DMR's technical specifications, network architecture, interfaces and protocols, and services are discussed. The standardization process and its future are also included.

7.1 INTRODUCTION

DMR is an open digital radio standard specified in European Telecommunications Standards Institute (ETSI) for Private Mobile Radio (PMR) users. DMR is used in Europe and several regions of the world as a low-cost, entry-level radio system for commercial and public safety use. The standard was first published in 2005 and has been widely adopted by radio manufacturers and users (mainly for commercial applications). Products built to the DMR standard also comply with the Federal Communications Commission (FCC) mandates in the USA for the use and certification of 12.5 kHz and 6.25 kHz narrowband technology [1]. DMR offers a replacement of

Fundamentals of Public Safety Networks and Critical Communications Systems: Technologies, Deployment, and Management,
First Edition. Mehmet Ulema.
© 2019 by The Institute of Electrical and Electronics Engineers, Inc. Published 2019 by John Wiley & Sons, Inc.

analog systems with all the benefits of a digital solution. DMR is more suitable for large rural areas with low traffic where simulcast/broadcast work best. The primary goal of DMR is to develop affordable digital systems with relatively low complexity. DMR provides voice, data, and other supplementary services [2].

The DMR protocol covers three modes of operations—Unlicensed (Tier I), licensed conventional (Tier II), and licensed trunked (Tier III) modes of operation.

- DMR Tier I is for devices used in the unlicensed spectrum. This part of the standard is developed for consumer applications (personal recreational use like amateur radio) and low-power commercial applications, using a maximum of 0.5 watt Radio Frequency (RF) power (for relatively small area coverage). This tier is not suitable for emergency public safety or large business applications.
- DMR Tier II is developed for licensed, conventional radio systems, mobiles, and hand portables. This part of the standard is developed for users who need spectral efficiency, advanced voice features, and integrated Internet Protocol (IP) data services in licensed bands for high-power communications.

   Tier II devices are specified and built primarily to replace analog radio systems with pin-to-pin compatibility and with the same coverage and the same application, including single site or multisite repeaters.

   Tier II specifies two-slot Time Division Multiple Access (TDMA) in 12.5 kHz channels.
- DMR Tier III is developed for trunked radio operations to support critical communications over a wide area. It builds on Tier II, which also specifies two-slot TDMA in 12.5 kHz channels. Tier III supports voice and short messaging handling similar to Terrestrial Trunked Radio (TETRA) with built-in 128 character status messaging and short messaging with up to 288 bits of data. DMR Tier III also supports packet data service in several formats, including IP version 4 (IPv4) and IP version 6 (IPv6) [3].

Some of the critical features of DMR are listed below:

- Air interface encryption
- One way authentication
- Short data messaging up to 40 characters
- Integrated voice + packet data
- Up to 100-watt base station
- 5-watt portable, 50-watt mobile
- Up to 1,200 users on a single site
- Up to 15 sites
- Semi-duplex telephone calls
- Manual geographic redundancy

TABLE 7.1. A List of Primary DMR Features

Functionality	DMR
Standards organization	ETSI
Channel access method	TDMA
Channel bandwidth	12.5 kHz
Modulation	4FSK
Raw data rate	9.6 Kbps
Spectral efficiency	0.768 bits/Hz
Number of time slots	2
Number of dedicated control channels	0 for Tier I, II; or 1 for Tier III
Direct mode	Tier I
Repeater (talk-through) mode	Tier II
Trunking mode	Tier III
Speech Codec	AMBE+2

- Subscriber roaming
- Priority call
- Analog fallback

Table 7.1 provides a list of several significant features of the DMR standards [4].

7.2 ARCHITECTURE

The architecture of a DMR network is relatively simple compared to other narrow-band technologies. Of course, the architecture depends heavily on the tiers.

Tier I architecture is mainly comprised of end-user devices (e.g. mobile stations), without the use of repeaters. There is no infrastructure, no telephone interconnects. In other words, Tier I architecture supports direct mode operation only (i.e. device-to-device). End-user devices contain fixed and typically integrated antennas. The air interface in Tier I is based on the Frequency Division Multiple Access (FDMA) mode.

The DMR standards documents do not provide an explicit architecture figure. Figure 7.1 is an attempt by the author to draw a generic reference architecture for DMR networks to be able to discuss the components and interfaces that may be involved in DMR Tier II and Tier III networks.

Tier II architecture may use one or more repeaters (without any trunking among them) to expand coverage areas as well as to accommodate more users.

Tier III architecture includes full trunking capabilities among the sites, as well as interfaces to telephone networks and other types of networks. Note that a Tier III network may contain a single site or multiple sites (up to 15). Furthermore, each site is

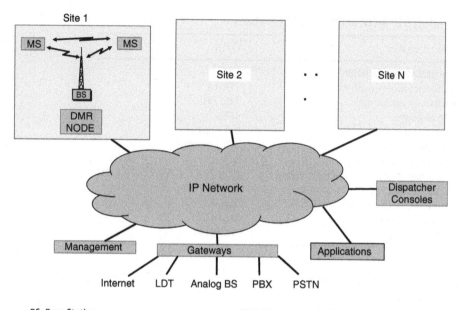

BS: Base Station   PBX: Private Branch Exchange
MS: Mobile Station   PSTN: Public Switched Telephone Network
LDT: Line Dispatch Terminal

Figure 7.1. DMR reference architecture.

capable of running independently, which is a critical capability to provide continuity of communication when the links to other sites fail.

Generically speaking, a DMR network contains the following components:

- **Subscriber units**—These are Mobile Stations (MS) and portable subscriber units, which are used by the end users to communicate with them or to communicate with other devices connected to the network.

- **Base stations**—Also called repeaters, Base Stations (BSs) facilitate radio communication among the subscriber units. These are the primary components of a DMR site, which may also include the antenna tower with radio receivers and transmitters, Ethernet hubs and switches, control and switching nodes, routers, and power supplies.

    The base station together with the other components in a DMR site is used to control and switch voice calls and data packets among the sites.

- **DMR node**—These nodes are responsible for managing the calls (i.e. call setup, call maintenance, call disconnection, collecting call related data, etc.) and transmission of the calls. A DMR node could act as a control node and a switching node. In a DMR network, there is only one control node, which is responsible for call management. Any node in the network is capable of operating as a switching node, which is responsible for the transmission of audio by switching the calls among users. In every node, there is also a controller, which provides packet switching and network management functions.

- **IP network**—The backbone of a DMR network, linking sites and gateways, servers, and applications by using IP-based networking routers. The use of an IP backbone makes it easier to employ the Internet and Web-based network management and other applications.

- **Gateways**—These gateways are used to connect a DMR network to devices and systems outside. For example:

  o A device (e.g. a line dispatch terminal) outside the network via an audio channel interface; the gateway converts Voice over IP (VoIP) calls to analog four-wire audio and vice versa

  o Analog base station(s); the gateway provides call control, switching, and full wire audio link functions

  o Public Switched Telephone Network (PSTN) and Private Branch Exchange (PBX), so that a DMR radio device can communicate directly with a voice telephone connected to public and private telephone networks; the gateway, in this case, is capable of converting IP voice to a transmission format suitable for PBX and PSTN

- **Dispatcher console**—This is a generic term used to apply to computer-based components that provide wide-ranging capabilities to communicate with all the users in the network to carry out dispatching operations.

- **Network management**—One or more computer-based systems to provide a variety of functions to operate, administer, maintain, and provision the network, its components, and its users. This includes configuration of the network elements and user devices, monitoring and reporting usage, performance, and fault records. Network management in DMR networks is widely available since it is based on the ubiquitous IP.

- **Application servers**—Since DMR networks use an IP-based core to interconnect its components, any IP-based application may be relatively easy to deploy. These applications may include web servers, email servers, database servers, and communications servers.

7.3 INTERFACES

As shown in Figure 7.1, DMR reference architecture includes air and wireline network interfaces. The air, trunking, and data protocols are specified in the DMR standards documents. There are a few other non-DMR standards and protocols used in DMR networks as well.

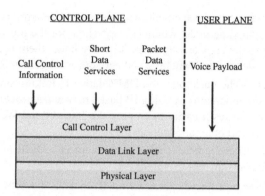

Figure 7.2. DMR air interface protocol stack.

7.3.1 DMR Air Interface (AI)

The DMR air interface and the protocols on it apply to the communications between BSs and the radio stations (mobile, portable, fixed) and between two radio stations for direct mode.

Figure 7.2 shows the protocol stack for DMR AI, which is specified in detail in [5].

The physical layer operates in 12.5 kHz bandwidth and uses TDMA technology with two time slots to create two logical channels, which are used to transmit voice and data (note that uplink and downlink communication paths require two different frequency bands). A logical channel can be designated as a *control* channel or as a *traffic* channel. As the name implies, a control channel is used for signaling and control—management of channel access, service requests, resources, registration, and provisioning of services. The control channel is also used to provide short data messaging service. Only one logical channel is needed to serve as the control channel in a DMR site; all the other logical channels are used as traffic channels.

The Data Link Layer (DLL) in DMR includes typical data link functions such as Medium Access Control (MAC), Logical Link Control (LLC), and error control. As shown in Figure 7.2, the protocols at this layer as well the layers above can either be a part of the *control* plane or the *user* plane. As commonly known, the user plane protocols are used to transport voice and data in circuit mode, whereas the control protocols are used for signaling to control and transmit data in packet mode. The sublayering of the DLL into MAC and LLC is left to the vendor's implementation in DMR.

Only the control plane contains Layer 3, which is called the call control layer in the DMR standards. The primary protocol at this layer is called the DMR Packet Data Protocol (PDP) and is used to control and manage the calls and facilities, as described in [6], and to provide short data and packet data services, as described in [7]. The data services and their features, shown in Table 7.1 and discussed in Section 7.4.2, are primarily realized by using the PDP.

7.3.2 Trunking Interface

The DMR standards in [8] specify the trunking protocol. However, it defines only the trunking related signaling (call control and channel allocations) procedures and messages on the air interface. The interfaces among the sites through the IP network are mainly implemented with vendor solutions.

Traffic channels and the appropriate resources required for the calls are dynamically assigned and released for each call request. In other words, channel allocation to a given call is done automatically without a manual selection of the channels.

As discussed in Section 7.3.1, one or two of the logical channels in each site are designated as the control channel, called the Trunk Station Control Channel (TSCC). Although some services (like short data services) are offered on the control channel, most other services are offered on traffic channels, which are allocated by the TSCC to form a pool of available channels.

The control channel may or may not be dedicated. The formal terminology for these types is "Dedicated Control Channel (DCC)" and "Composite Control Channel (CCC)," respectively. As the name implies, a DCC is a logical channel that performs control channel functions continually, whereas a logical channel designated as a CCC provides control channel functions as well, but may revert to a traffic channel when needed (for example, when there is no traffic channel available).

7.3.3 Data Application Interface

This interface allows the dispatch consoles to communicate with the site equipment (MSs, BSs, repeaters, nodes). This is a wired interface based on widely available and well-known Ethernet and IP interfaces and protocols, which also allows remote login to the dispatch consoles.

This interface, which is not an ETSI standard, was approved by the DMR Association [9]. The protocol known as Application Interface Specification (AIS) is mostly a collection of open specifications that define the Application Programming Interfaces (APIs) [10]. It was developed by the Service Availability Forum (SA Forum) [11].

7.4 SERVICES

The services offered by a DMR network can be categorized under voice services and data services, as discussed below. Table 7.2 shows a list of services offered by DMR networks.

TABLE 7.2. Summary of DMR Services

Services		Supplementary Services
Voice	Individual call	Late entry
		Open voice channel mode call
		Talking party identification
	Group call	Late entry
		Unaddressed voice call
		Open voice channel mode call
		Talking party identification
	Conference call	Late entry
		Talking party identification
	Broadcast call	Late entry
		Talking party identification
Data	Confirmed packet data	
	Confirmed short data—status	
	Confirmed short data—text	
	Confirmed short data—GPS data	
	Unconfirmed packet data	
	Unconfirmed short data—text	
	Unconfirmed short data—GPS data	

7.4.1 Voice Services

DMR voice services include individual voice calls and group voice calls.

Individual calls are the traditional, private voice calls between two end users, although an individual call could be between an end user and the dispatcher. A voice call between a DMR user and a PSTN user through the gateways is also possible.

Group calls are voice calls among a predetermined group of users with a common interest. A group call, the most widely used service among DMR users, can be arranged by the network administrator or can be formed in an ad hoc fashion on-site when needed. A group call can be of a conference call type or a broadcast type. In the conference type, called *all call* in DMR, any user in the group can call the group and everyone in the group can listen, respond, and communicate. In a broadcast call, however, only the one making the call talks and the others in the group must only listen; no response is allowed in a broadcast call. Furthermore, a group may be:

- selectable, where it can be selected by a user to join the call
- subscribable, where the user can subscribe it before the call

- scannable, where it can be used by its members and others to listen to the calls taking place in this group

Some groups engaged in high priority situations can be configured permanently into the end-user device radio, which responds to any call in this group automatically.

Supplementary services offered by DMR networks may include some additional features, such as:

- *Open Voice Channel Mode (OVCM)*—to allow the group to be selectable and scannable,
- *Unaddressed Voice Call*—to allow the users to customize the end-user devices, for example, to use unique ringtones or notification tones,
- *Late Entry*—to allow a user to join a group call already in progress, and
- *Talking Party Identification*—to allow the display of the identification of the calling party on the called users' devices.

7.4.2 Data Services

DMR offers two kinds of data services—Packet Data Services (PDS) and Short Data Services (SDS).

Data in DMR can be transmitted as confirmed or unconfirmed. As the name implies, the transmission of confirmed data requires a positive or negative acknowledgment (confirmation) from the destination. If the acknowledgment is negative, then the same data block is retransmitted. The transmission of unconfirmed data does not require an acknowledgment from the destination, similar to datagrams.

IP is used to provide variable length **packet data services**, which can be used to transmit large amounts of information. IPv4 and IPv6 are supported. Packet data, also called *IP data*, services are provided on a *traffic channel*, even though setting up this data call is done on the control channel.

On the other hand, **short data services** are used to transmit short messages, on the *control channel*, up to 626 bytes long, or up to 1130 bytes long. Typically, this type of messages contains text, status information, or Global Positioning System (GPS) data.

Although this service is available for user to user and group communications, a more common use is for communication between dispatchers and users, or between application servers and end users and perhaps end-user devices. For example, the dispatcher is sending a short message to a specific user or a specific group. Another example could be standardized digitally coded status labels sent to individuals or groups. Transmitting GPS data is yet one more example for short data services. Upon request, typically from application servers, such as fleet management systems, end-user devices (equipped with GPS receivers) transmit GPS data, which may be used for tracking, inventory and asset management, as well as workforce management.

7.5 OPERATIONS

Like Project 25 and TETRA, DMR also supports conventional and trunked operations (see Chapter 4 for a brief generic discussion of these options).

As discussed in Section 7.1, DMR Tier II is developed for licensed, conventional radio systems, whereas DMR Tier III is developed for trunked radio operations (remember also that DMR Tier I is mainly for consumer applications [e.g. amateur radio] and low-power commercial applications).

DMR Tier II is used primarily to replace analog conventional radio systems with the same coverage and the same applications, including single site or multisite repeaters. In this mode, DMR can act as a simulcast network with the voting system, where all the repeaters operate on the same frequency at the same time, and the system chooses the best cell for the end user while moving.

DMR Tier III is mainly used for trunked radio operations to support critical communications over a wide area.

7.6 SECURITY

DMR provides privacy and authentication features by activating the encryption mode in user radio terminals. We can list three different types of privacy in DMR:

- **Basic privacy**—This is not encryption-based privacy. The way it works is that, before the call, a key between 1 and 255 is chosen to scramble an available frequency, which is used between the radios for the call. Naturally, this can be quickly uncovered by an eavesdropper by merely trying all 255 values.
- **Enhanced privacy**—This is a more secure privacy feature that is based on a user-selected 40-bit encryption key. In this scheme, the customized encryption key must be entered into the device along with a corresponding name and number slot assignment.
- **Top-level encryption**—This is a much more secure privacy scheme, which requires a license and specific software to use the Advanced Encryption Standard (AES) with a 256-bit key length.

DMR requires the authentication of the user devices during the registration process. Authentication may also be required during call setup requests, which may result in rejection of the call by the network. Additionally, a user device may authenticate the network as well.

To authenticate a user device, DMR uses a well-known encryption algorithm based on Rivest Cipher 4 (RC4) with a 56-bit authentication key [12]. During the authentication process, when a user device receives a challenge from a node, a response is prepared by using a built-in fixed 56-bit key and transmitted to the challenging node. Upon receiving the response, the node executes the same algorithm by using the information about this end device in its database and compares its result with the response it received. If they match, the end device is determined to be genuine.

7.7 SPECTRUM

The DMR standards cover a wide range of the frequency spectrum, from 30 MHz all the way up to 1 GHz. However, regional regulations and vendor implementations may vary. Depending on the three modes of operation in DMR, the following spectrum usage seems to be more common:

- DMR Tier I is for devices used in the unlicensed spectrum in the European 446 MHz band. In the USA, the 446 MHz range is used primarily by the US Government with the amateur radio service a heavy secondary user.
- DMR Tier II is developed for licensed, conventional radio systems, mobiles, and hand portables operating in PMR frequency bands from 66 to 960 MHz.
- DMR Tier III is developed for trunked radio operations in frequency bands from 66 to 960 MHz.

The DMR protocol is flexible enough to allow simplex and duplex modes of communication on a single frequency or two different frequencies. Some further combinations are also possible, as shown in Table 7.3. This flexibility makes it possible to deploy DMR in all frequency bands from 66 MHz to 1 GHz, depending on the regional regulatory restrictions.

7.8 STANDARDIZATION

The DMR standards are relatively more recent than Project 25 and TETRA. ETSI published the first DMR standard specification in 2006. The motivation behind DMR was to develop a cost-effective system with backward compatibility with the existing analog systems. Motorola played an essential role in the specification activities and was the first actor in this new segment of the PMR market [4].

The standard allows DMR manufacturers to implement additional features on top of the standards. However, initially this created problems in interoperability among equipment built by different vendors. In 2009, about 40 manufacturers set up the DMR Association to work on interoperability among vendors' equipment and to promote the DMR standard. Since 2010, the companies involved have been conducting interoperability testing, publishing their results on the DMR Association web site.

TABLE 7.3. Frequency and Communication Mode Possibilities in DMR

Communications Mode	Call Type	Through a Repeater		Peer-to-Peer	
		Single Frequency	Dual Frequency	Single Frequency	Dual Frequency
Simplex	Individual	Yes	Yes	Yes	Yes (allows two communication streams between same source and destination)
	Group	Yes	Yes	Yes	Yes (allows two communication streams)
Duplex	Individual	No	Yes	Yes	Yes (two streams possible with the use of RF duplexer)
	Group	Not sufficient to support conferencing			

DMR specifications are documented in several ETSI standards:

- ETSI TS 102 361-1: the DMR air interface protocol [5]
- ETSI TS 102 361-2: DMR voice and generic services and facilities [6]
- ETSI TS 102 361-3: the DMR data protocol [7]
- ETSI TS 102 361-4: the DMR trunking protocol [8]
- ETSI TR 102 398: DMR general system design [13]

Reference [13] is primarily a designer's guide that is intended to apply to all the elements specified in all four parts of the DMR standards. The DMR standards documents are open standards, which means that they are freely available from the ETSI website [14].

7.9 DEPLOYMENT

As discussed above, DMR technology is relatively new and more suitable for relatively small networks and private applications. Therefore, it is not as widely deployed as Project 25 and TETRA.

The DMR standard attracted several companies to invest in developing products based on this standard. As of this writing, Motorola Solutions, Selex Communications, HYT, Funkwerk, and Radio-Activity Solutions offer commercial DMR products, and Tait Communications, Team Simoco, and Vertex Standard plan to offer DMR equipment. The following is a list several other companies that are involved in the DMR market: Aselsan, Avtec Inc., CML Microcircuits, EMC, etherstack, Fylde Micro, Harris, Hytera, ICOM, JVC KENWOOD, KIRISUN, Larimart, RADIO-DATA GmbH, TECHBOARD, VictelGlobal, and ZTE [15].

DMR was developed to cover wide areas with low traffic density. Therefore, it requires fewer cell sites (i.e. costs less) to cover the same area, compared to TETRA and Project 25. A medium-size TETRA system may cost 3–5 times more than a DMR one. DMR does not do that well in optimizing the traffic within a cell. It may be more expensive to increase the capacity to handle more traffic in a given cell. Therefore, DMR is more cost-effective when it is used to cover large areas with relatively low traffic [4].

Since DMR includes the analog mode (in addition to the digital mode), it is less costly to adopt it and less costly to migrate to this technology.

Compared to analog systems, both P25 and DMR require a slightly greater number of sites to cover the same area covered by the existing analog technology.

7.10 FUTURE

There is almost no information available as to the future of DMR technology and its continuing standardization. However, it is safe to speculate that due to its niche market, especially in commercial sectors, DMR standardization will continue with perhaps some additional data capabilities.

7.11 SUMMARY AND CONCLUSIONS

This chapter provided an overview of DMR technology, its specifications, network architecture, interfaces and protocols, and services.

DMR offers a replacement for analog systems with all the benefits of a digital solution. DMR is uniquely suited to large areas with relatively low traffic applications where simulcast gives the best performance. The DMR protocol covers three modes of operation—unlicensed (Tier I), supporting only device-to-device communications, licensed conventional (Tier II), with repeaters only, and licensed trunked (Tier III) modes of operation.

DMR is specified in ETSI and used in Europe and several regions of the world. DMR is a low-cost, entry-level radio system for commercial and public safety use.

REFERENCES

1. "DMR—The RadioReference Wiki." [Online]. Available: https://wiki.radioreference.com/index.php/DMR. [Accessed: Apr. 16, 2018].
2. Digital Mobile Radio Association, "The DMR Standard," Dec. 29, 2010. [Online]. Available: http://dmrassociation.org/the-dmr-standard/. [Accessed: Jul. 24, 2016].
3. R. Marengon. (2010, April 1). A TETRA and DMR comparison. *Radio Resource Mag.* [Online], Apr. 2010. Available: https://www.rrmediagroup.com/Features/Features Details/FID/174.
4. B. Bouwers, "Comparison of TETRA and DMR," White Paper, Rohill Technologies, 2010.
5. "ETSI TS 102 361-1 Part 1: DMR Air Interface (AI) protocol." [Online]. Available: http://www.etsi.org/deliver/etsi_ts/102300_102399/10236101/02.04.01_60/ts_10236101 v020401p.pdf. [Accessed: Jul. 10, 2017].
6. "ETSI TS 102 361-2 Part 2: DMR voice and generic services and facilities." [Online]. Available: http://www.etsi.org/deliver/etsi_ts/102300_102399/10236102/02.03.01_60/ts_ 10236102v020301p.pdf. [Accessed: Jul. 10, 2017].
7. "ETSI TS 102 361-3 Part 3: DMR data protocol." [Online]. Available: http://www.etsi .org/deliver/etsi_ts/102300_102399/10236103/01.02.01_60/ts_10236103v010201p.pdf. [Accessed: Jul. 10, 2017].
8. "ETSI TS 102 361-4 Part 4: DMR trunking protocol." [Online]. Available: http://www.etsi .org/deliver/etsi_ts/102300_102399/10236104/01.08.01_60/ts_10236104v010801p.pdf. [Accessed: Jul. 10, 2017].
9. Catalyst Communications Technology, "Understanding AIS—The DMR application inter-face specification," The Dispatch Blog, Dec. 23, 2016. Available: https://blog.catcomtec .com/2016/12/23/understanding-ais-the-dmr-application-interface-specification/.
10. "Application Interface Specification," Wikipedia. [Online]. Jul. 8, 2017. Available: https://en.wikipedia.org/wiki/Application_Interface_Specification.
11. Service Availability Forum, "Service Availability Forum: Application interface spec-ification." [Online]. Available: http://www.saforum.org/page/16627~217404/Service-Availability-Forum-Application-Interface-Specification. [Accessed: Jul. 22, 2017].
12. "RC4," Wikipedia. [Online]. Jul. 2, 2017. Available: https://en.wikipedia.org/wiki/RC4.
13. "ETSI TR 102 398: DMR general system design." [Online]. Available: http://www.etsi .org/deliver/etsi_tr/102300_102399/102398/01.03.01_60/tr_102398v010301p.pdf. [Accessed: Jul. 10, 2017].
14. "ETSI DMR flyer." [Online]. Available: http://www.etsi.org/website/document/techno logies/leaflets/digitalmobilradio.pdf. [Accessed: Jul. 10, 2017].
15. Digital Mobile Radio Association, "DMR Association—Members product showcase." [Online]. Available: https://dmrassociation.org/product-showcase.html. [Accessed: Apr. 16, 2018].

8

LONG-TERM EVOLUTION (LTE)

This chapter presents Long-Term Evolution (LTE), LTE Advanced (LTE-A), and LTE-A Pro technologies including technical specifications, network architecture, interfaces and protocols, and services with an emphasis on their use in critical communications systems. Many countries and agencies around the world are planning to use LTE for public safety networks.

LTE technology, in general, has been covered in great detail in the literature. There are already numerous books, journal and magazine articles, and other publications. Our aim in this book is to provide a brief introduction to LTE and focus on the aspects related to critical communications systems.

Note that although there are some subtle differences between LTE and LTE-A, this book uses the term LTE only to refer to both, and discusses the differences when appropriate.

The standardization process and its future are also discussed. Appendix C provides a comprehensive list of critical communications related LTE standards documents.

Fundamentals of Public Safety Networks and Critical Communications Systems: Technologies, Deployment, and Management,
First Edition. Mehmet Ulema.
© 2019 by The Institute of Electrical and Electronics Engineers, Inc. Published 2019 by John Wiley & Sons, Inc.

8.1 INTRODUCTION

LTE is a widely accepted technology for mobile broadband communications [1–5]. In general, LTE is an evolution of 2nd Generation (2G) Global System for Mobile communications (GSM) and 3rd Generation (3G) Wideband Code Division Multiple Access (W⁻CDMA) technologies. As defined in 3rd Generation Partnership Project (3GPP) Release 8, LTE offers several significant benefits compared to 3G, including reduced latency, higher user data rates, improved system capacity, and better coverage. In addition to the improvements in radio access technologies, packet core networks have also evolved into a flat, all Internet Protocol (IP)-based architecture defined by 3GPP. The aim of LTE's all IP-based core infrastructure is to support any IP-based service and optimize network performance [6, 7].

LTE has been considered as a significant first step in responding to the ever-increasing demand for large capacity multimedia and high-speed mobile data services. However, wireless specialists are calling LTE-A as the "true 4G" because unlike ordinary LTE, it is considered as meeting the 4G system requirements (such as higher speed) set by the ITU [8]. Note that LTE and LTE-A are forward and backward compatible with each other. This means that any LTE device can operate in an LTE-A network, and likewise, any LTE-A device can operate in an LTE network.

3GPP Release 13 introduced LTE Advanced Pro (LTE-A Pro), which adds some new features including the ones related to public safety, Device to Device (D2D), and Proximity Services (ProSe). Other enhancements in LTE-A Pro include indoor positioning, carrier aggregation, interoperability with Wi-Fi, and massive Multiple-Input/Multiple-Output (MIMO). Many in the industry consider LTE-A Pro as a stepping stone toward 5G [9, 10].

Primary LTE parameters—Spectrum flexibility is one of the critical features of LTE, which can be deployed with different bandwidths ranging from 1.4 MHz up to 20 MHz (up to 100 MHz in LTE-A) and supports both Frequency-Division Duplex (FDD) and Time-Division Duplex (TDD) operations. LTE uses Orthogonal Frequency Division Multiplexing (OFDMA) on the downlink to meet the demands on high peak data rate and high user mobility. Single Carrier FDMA (SC-FDMA) is used on the uplink. LTE provides high system performance including peak data rates exceeding 300 Mbps (1 Gbps in LTE-A) and reduced latency in less than 10 milliseconds (round-trip times between user equipment and the base station). LTE-A can achieve data download rates as high as 3 Gbps and upload rates as high as 1.5 Gbps [6] (note that these are theoretical numbers; the actual numbers are much lower).

LTE-A also includes new transmission protocols and a multiple-antenna scheme called MIMO, which enables smoother handoffs between cells, increased throughput at cell edges, and more bits per second in each hertz of the spectrum (i.e better spectral efficiency).

Carrier aggregation, one of LTE-A's capabilities, allows operators to combine their disjointed narrower channels into one broad channel. This feature increases the bandwidth available to a mobile device by combining frequency channels that reside in different parts of the spectrum, resulting in significant performance gain [11].

An upgrade from LTE to LTE-A requires both a software and hardware upgrade. Concerning the enhanced Node Base station (eNodeB) design, a hardware upgrade or an additional radio unit is required if a full 4 × 4 MIMO capability is to be implemented. Carrier aggregation poses new challenges in the implementation of power amplifiers, especially for wide area base stations. For this feature, operators need to upgrade with additional hardware to acquire new spectrum that the current hardware is capable of handling. Software upgrades for L2/L3 layers are needed as well.

Features such as self-configuration and self-optimization, reducing the cost of network roll-out and operation, are also supported. Table 8.1 below summarizes the primary LTE parameters.

As discussed in this section, 3GPP continues to enhance and evolve LTE technology by improving the existing features and adding new features. LTE-A has been covered in Releases 9 through 12. However, with Release 13, 3GPP started using a new marker, called LTE-Advanced Pro (LTE-A Pro), to refer to the latest version of LTE

TABLE 8.1. LTE Primary Parameters

	LTE (Release 8)	LTE-A (Release 9 and Beyond)
Peak data rate—DL	• 300 Mbps (4 × 4 MIMO)	• 1 Gbps
Peak data rate—UL	• 75 Mbps (64 QAM)	• 500 Mbps in 15 bps/Hz
Peak spectral efficiency—DL	• 16.3 b/s/Hz (4 × 4 MIMO)	• 30 b/s/Hz (up to 8 × 8 MIMO)
Peak spectral efficiency—UL	• 4.32 b/s/Hz (64 QAM SISO)	• 15 b/s/Hz (up to 4 × 4 MIMO)
Cell range	• 5 km—optimal size • 30 km sizes with reasonable performance • Up to 100 km cell sizes supported with acceptable performance	
Cell capacity	• Up to 200 active users per cell (5 MHz) (i.e. 200 active data clients)	
Mobility	• Optimized for low mobility (0–15 km/h) but supports high speed	
Transmission bandwidth (MHz)	• ≤20	• ≤100
Radio access	• OFDMA for DL data transmission • SC-FDMA for UL transmission	• OFDMA for DL data transmission • SC-FDMA for UL transmission

(Continued)

TABLE 8.1. *(Continued)*

	LTE (Release 8)	LTE-A (Release 9 and Beyond)
Latency (delay)	• User plane <5 ms • Control plane <50 ms	• From idle to connected in less than 50 ms and then shorter than 5 ms one way for individual packet transmission
Channel bandwidth	• 1.4, 3, 5, 10, 15, 20	
Modulation types	• QPSK, 16 QAM, 64 QAM (UL and DL)	
Duplex schemes	• FDD and TDD	
Carrier aggregation	• No	• Yes (up to five of these "component carriers," each offering up to 20 MHz of bandwidth, can be combined, which creates a maximum aggregated data pipe up to 100 MHz)
MIMO	• Some (only for DL) four transmitters	• Yes (up to eight antenna pairs for the DL and up to four pairs for the UL)
Relay node	• Yes	• Yes—more advanced (first decode the transmissions and then forward only those destined for the mobile units that each relay is serving)
Coordinated multipoint	• Yes (ICIC)—optional	• Yes—eICIC
Device to device	• No	• Yes

b/s/Hz	bits per second per Hertz	Mbps	Megabits per second
bps	bits per second	MHz	MegaHertz
DL	Downlink	MIMO	Multiple Input Multiple Output
eICIC	enhanced Intercell Interference Coordination	ms	millisecond
		OFDMA	Orthogonal Frequency Division Multiple Access
FDD	Frequency Division Duplex		
Gbps	Gigabits per second	SC-FDMA	Single Carrier FDMA
ICIC	Intercell Interference Coordination	SISO	Single Input Single Output
km	kilometer	QPSK	Quadrature Phase Shift Keying
km/h	km per hour	QAM	Quadrature Amplitude Modulation
LTE	Long Term Evolution	TDD	Time Division Duplex
LTE-A	LTE-Advanced	UL	Uplink

technology. LTE-A Pro aims to accelerate the introduction of new services related to the IoT and 5G. The following is a list of some of the critical features of LTE-A Pro:

- Data rates: beyond 3 Gbps
- Latency: 2 ms
- Carrier bandwidth: 640 MHz
- Carrier aggregation: up to 32 different carriers
- MIMO antennas: up to 64 antennas
- Battery life: 10-fold increase (compared to LTE)

LTE-A Pro also includes enhancement of the existing features and new features specifically for public safety and mission-critical applications. The features in this category include the Mission-Critical Push-To-Talk (MC-PTT) protocol, mission-critical video/data services, as well as the LTE Direct (LTE-D) protocol to be used in providing ProSe, where thousands of devices within 500 meters range can communicate directly.

8.2 ARCHITECTURE

The overall network architecture of LTE can be discussed under two areas—the access network and the core network. The Evolved–Universal Mobile Telecommunications System (UMTS) Terrestrial Radio Access Network (E-UTRAN) consists of base stations, called evolved Node Bs (eNodeB or eNB). The Evolved Packet Core (EPC) consists of nodes that are connected to each other through standardized interfaces, as shown in Figure 8.1.

Note that the components discussed here are functional components. This means that vendors may choose to develop physical products that may incorporate one or more of these functional components.

8.2.1 E-UTRAN

E-UTRAN contains eNBs. The air interface between a User Equipment (UE) and an eNB is called Evolved UMTS Terrestrial Radio Access (E-UTRA). In E-UTRAN, there is no centralized and separate radio network controller. This functionality is integrated into eNB. This simplifies the architecture, makes the network more reliable, and allows faster response times.

UE is a generic name for any device that allows the end user to communicate through the LTE networks. A generally known UE is the cell phone or the smartphone, as it is more often called these days. A tablet, a notebook, or a laptop computer with a wireless interface to the LTE network are also considered to be UE.

As Figure 8.1 shows, a UE may be connected to the LTE network via an eNB, a Relay Node (RN), or a Home eNB (HeNB). The UE is responsible for initiating

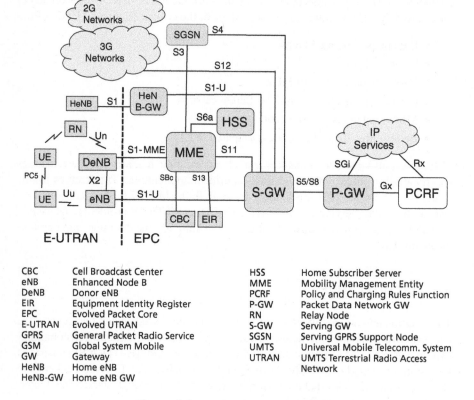

Figure 8.1. LTE functional architecture.

and terminating calls. Also, they handle the management of identities and mobility management related tasks in cooperation with the other components in the network.

Based on their performance and capability levels, UE are classified into 12 different categories—Category 1 to Category 12 [12]. The lower the level, the weaker the performance and capabilities. This category level information helps the eNBs and MMEs to manage communications with the UE much more effectively (note that LTE assigns different categories based on downlink and uplink characteristics. The 12 categories mentioned above are based on combined uplink and downlink capabilities).

eNodeB or **eNB** is responsible for managing one or more cells. This includes all radio-related functions such as:

- Radio Resource Management (RRM), which includes radio bearer control and radio admission control, dynamic allocation, modification, and release of resources to UEs
- IP header compression for efficient use of the radio interface

- Connectivity to EPC, which includes the selection of MME at UE attachment and the radio bearer to the S-GW
- Encryption of all user data sent over the radio interface
- Scheduling and transmission of broadcast information (originated from the MME or Operations, Administration, and Maintenance [OAM])
- Measurements on the radio environment and eNB conditions and measurement reporting
- Physical layer functionality such as scrambling, beamforming, OFDM modulation, and link adaption

RN, introduced with LTE-A, is a fixed, low powered, relay eNB without a wired connection to any eNBs or any core network component. The RN, as the name implies, relays messages between the eNB and UE. Although the relaying concept is not new, LTE RNs are mainly intended to increase the performance at the edges of cells. LTE RNs are used to increase network density, extend network coverage, and roll out the network efficiently and rapidly.

Donor eNB (DeNB) is an enhanced eNB that can also support RNs, in addition to supporting its own UE. The DeNB acts like an MME, an eNB, and an S-GW to the RNs.

HeNB is an optimized eNB for use in smaller cells that can be set up at homes or in public places. Depending on the power level, the capacity, and the coverage, these small cells may be called femtocells, picocells, or metrocells, sixes ranging from small to large.

8.2.2 Evolved Packet Core (EPC)

EPC is responsible for the overall control of the network and its components to facilitate data and voice services to its users. EPC consists of the following functional entities:

Serving Gateway (S-GW) is used to manage user plane mobility and also to maintain the data paths between the eNBs and the Packet Data Network Gateway (P-GW) by forwarding and routing user data packets, providing the mobile anchoring function facilitating the interworking with 2G and 3G networks.

Mobility Management Entity (MME) is responsible for the functions related to connection management and the functions related to bearer management. It is a critical control node in LTE. It is responsible for controlling the signaling process, maintaining session states, managing authentication, and mobility with LTE, 3G, and 2G nodes. The other key features of the MME are idle mode UE tracking, paging procedures, bearer activation/deactivation, choosing of an S-GW for the UE, interacting with the Home Subscriber Server (HSS) to authenticate the user on attachment, and providing temporary identities for the UE.

The MME may be connected to a **Cell Broadcast Centre (CBC),** an information distribution server used to broadcast messages to all the users connected to eNBs attached to the MME. This is an important feature that is expected to be used heavily in critical communications applications.

The MME may also be connected to the **Equipment Identity Register (EIR)**, a database that contains the status of the mobile devices registered on the network, which is used to authenticate UE.

P-GW acts as the interface between LTE network and other packet data networks (PDNs) by being the point of exit and entry of traffic for the UE. The P-GW allocates IP addresses to UE; performs policy enforcement (based on the rules in the Policy Control and Charging Rules Function [PCRF]), deep-packet inspection, and filtering of packets; provides charging support; and manages the quality of service (QoS). It also acts as the mobility anchor for interworking between 3GPP and non-3GPP technologies such as WiMAX and CDMA.

HSS is a central database that contains users' subscription data. The HSS includes functionalities such as mobility management, user authentication and access, service authorization, service profile, and user security. The HSS also maintains data about PDNs that the UE may be subscribed to.

PCRF is responsible for policy enforcement and flow-based charging. The PCRF also provides QoS policy information to the P-GW and manages data sessions based on the subscribers' profiles.

HeNB Gateway (HeNB-GW) is used to aggregate and concentrate traffic from a large group of small cells (HeNBs) to S-GW and to enable seamless communication for users as they move in and out among HeNB and eNB cells. Security is the primary concern here because some of these cells may not be part of a trusted network.

Serving GPRS Support Node (SGSN) is not a component of the EPC. It is shown in Figure 8.1 to illustrate the interworking of the LTE network with a General Packet Radio Service (GPRS) network (note that GPRS provides packet data exchanges in GSM-based 2G and 3G networks).

8.3 INTERFACES

Figure 8.1 shows the reference points among the functional entities. In general, a reference point is a demarcation point between two functional entities, and an interface is a physical implementation that incorporates one or more reference points. However, in both standards specifications and literature, this distinction has not been followed to the letter. Instead, the words "reference point" and "interface" have been used interchangeably. In this chapter, we use the word "interface" to refer to the lines among the boxes shown in Figure 8.1.

In the following sections, LTE interfaces are discussed under four different areas—air interfaces, E-UTRAN interfaces, EPC interfaces, and the interfaces to non-LTE networks. However, first, the protocol stacks on significant interfaces are presented.

As shown in Figure 8.2, the protocol stack on the user plane differs slightly from the stack on the control plane. In both planes, the air interface uses the same bottom four protocols, namely Packet Data Convergence Protocol (PDCP), Radio Link Control (RLC), LTE Layer (L1) protocol, and LTE Medium Access Control (MAC) protocol. The control plane adds Radio Resource Control (RRC) on top of PDCP. Furthermore, the Non-Access Stratum (NAS) above the RRC provides connection and session management related functions between the UE and the MME, and S1 Application Protocol (S1-AP) provides signaling functions between the eNB and the MME. On the user plane, application units are enveloped by one or more IP packets and transported to the destination gateway transparently over the eNB and S-GW.

On the UTRAN and EPC side, depending on the applications, different L1, L2, and tunneling protocols such as GPRS Tunneling Protocol-User Plane (GTP-U) may be used. Also, the Stream Control Transmission Protocol (SCTP), a transport layer protocol, is used to handle the communication between the eNB and the MME.

8.3.1 Air Interface

The interface between the UE and the eNB is formally called **LTE Uu**. It is the wireless interface where bits are transmitted via coded, modulated, encrypted electromagnetic signals at the physical layer. The other protocols running on this interface, shown in Figure 8.2, have been discussed previously. These protocols are known as the "AS protocols."

Un, the interface between an RN and its DeNB, is a modified version of Uu.

PC5 is a relatively new air interface between two UEs. It was introduced in Release 12 to provide "Proximity-based Services," called **ProSe** in short, which was developed in response to the demand from critical communications stakeholders. Thanks to this service, the UE can communicate directly without the eNBs, which may be out of service due to natural and manmade disasters.

8.3.2 E-UTRAN Network Interfaces

These interfaces may be discussed under two categories. One category includes the interfaces used to connect end devices; the other category includes the interfaces used to connect the eNBs to the EPC components.

X2 interface is used to connect the ends. The protocol stack on this interface is the same as the one on the S-MME interface, except the application layer protocol on X2 is X2-Application Protocol (X2-AP).

S1 interface is used to connect the eNBs to various components in the core network. Therefore, depending on the component that the eNB is connected to, the S1 takes on an additional identifier, as discussed below:

- **S1-MME** is used to connect eNBs and MME to control and manage communications
- **S1-U** is used to connect eNBs and S-GW to facilitate user plane functions such as bearer tunneling.

Control plane

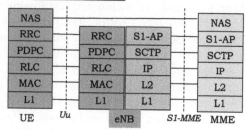

User plane

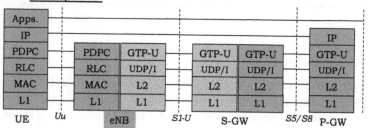

eNB	Enhanced Node B		PDPC	Packet Data Convergence
E-UTRAN	Evolved UTRAN			Protocol
GTP-U	GPRS Tunneling Protocol-User		RRC	Radio Resource Control
	Plane		RLC	Radio Link Control
GW	Gateway		S1-AP	S1 Application Protocol
IP	Internet Protocol		S-GW	Serving GW
L1	Layer 1		SCTP	Stream Control Transmission
L2	Layer 2			Protocol
MAC	Medium Access Control		UDP	User Datagram Protocol
MME	Mobility Management Entity		UTRAN	UMTS Terrestrial Radio Access
NAS	Non-Access Stratum			Network
P-GW	Packet Data Network GW			

Figure 8.2. LTE protocol stacks.

Protocols used on S1-MME and S1-U are shown in Figure 8.2 and discussed in the previous sections. Note that the application protocol, S1-AP, on the S1-MME interface uses the SCTP, a reliable transport protocol whereas the application protocol, GTP-U, on the S1-U interface uses the user datagram protocol as the transport protocol.

8.3.3 EPC Interfaces

The interfaces in this category are used to exchange signaling and user information among the components of a core network, as discussed below:

- **S5** is used to connect S-GW and P-GW to provide tunnel management functions. **S8** is also used between S-GW and P-GW. They are essentially the same

interface, except that while S5 is internal to an LTE network, S8 is an internetwork interface, used during roaming among different networks. In other words, S8 provides communications between S-GW in the visited network and P-GW in the host network.

- **S6a** is the interface between MME and HSS and is used to transfer subscription and authentication data to be used in the authentication of the user's access.
- **S9** is used between the visited PCRF and the home PCRF to transfer policy and charging information.
- **S10** is used between MMEs to exchange signaling information among them, mostly during handovers.
- **S11** is used to connect the MME to S-GWs (an MME can support multiple S-GWs). This interface is used to manage the establishment of bearer channels in the core network.
- **S13** connects the MME to the EIR to exchange information that is used in checking the identity of the UE. The protocol called *Diameter* is used for this purpose at the application layer. At the transport layer, the SCTP is used.
- **Gx** is the interface between the P-GW and the PCRF to transfer policy and charging rules from PCRF to the P-GW.
- **SBc** is between CBC and MME for broadcasting warning message delivery and control functions. The SBc Application Protocol (SBc-AP) on this interface uses the SCTP for transport of its messages.

8.3.4 Interworking Interfaces

The interfaces in this category are used to facilitate communications between the EPC components and the components belonging to external networks, as discussed below:

- **S3** is used for interworking with GPRS-based 2G and 3G GSM networks (standardized by 3GPP). Through this interface between the MME and SGSN, an LTE user can exchange information (user and bearer data) with a non-LTE user on a 3GPP compliant network.
- **S4** is similar to S3, except that it is between SGSN and the S-GW and provides the exchange of signaling and control information for this type of interworking between an LTE user and a non-LTE user.
- **SGi** is used to connect the P-GW to a packet data network, which could be internal or external to the LTE operator.
- **Rx** is used to connect the PCRF to a packet data network.
- **S12** is used between the P-GW and a 3G network for transferring user traffic when a direct user plane tunnel is established. The configuration of this interface depends on the operator.

8.4 SERVICES

LTE supports all types of mixed data, voice, video, and messaging services.

As discussed in previous sections, LTE is an all IP network employing packet switching technology to move information among its users. Therefore, it can be said that it is more suitable for providing broadband-rate data services including interactive, real-time, rich multimedia communication services (e.g. video streaming, visual communications) and other communication services such as messaging, collaboration, and conferencing. The end user devices, UE, capable of handling data services on the go (as well as at home) include tablets, notebooks, and laptops with LTE access interfaces.

Support for toll-grade, carrier-provided voice services over an all-IP LTE packet-switched network has been a challenge. Voice calls in earlier 2G and 3G networks were handled via the circuit-switched method. Therefore, LTE networks need to be augmented with additional capabilities to provide voice services. Vendors and operators have taken several interim approaches to address this challenge [13].

- **Circuit-switched fallback (CSFB)** has been used extensively to provide voice services quickly by using 2G and 3G networks. This way the LTE network is used for providing data services only. This solution is an inexpensive one but introduces longer call setup delays.
- **Simultaneous Voice and LTE (SVLTE)** is a UE-based solution. More specifically, the UE has both circuit-switched modes as well as packet-switched modes. As in the CSFB case, the LTE network is used for providing data services only. For voice services, the UE uses the circuit-switched mode to access 2G or 3G networks directly. No additional capability is needed in the LTE network. This approach eliminates the high delay in call setup in the CSFB approach. However, the UE will be more expensive.
- **Single Radio Voice Call Continuity (SRVCC)** is a user-initiated approach using Voice over IP (VoIP) applications such as Skype on their UE. In this approach, the network does not get involved in voice communication directly. These voice calls are merely treated as data calls.
- **Voice over LTE (VoLTE)** is similar to VoIP, where the voice is packetized and is sent over LTE (remember that LTE is an all IP network). Unlike VoIP, VoLTE has much higher, carrier-grade QoS requirements because it is offered as part of a paid service to the subscribers. Also, VoLTE requires the use of the Adaptive Multi-Rate Wideband codec, which supports 16 kHz sampling, which is called High Definition (HD) voice [8].

The IP Multimedia Subsystem (IMS), an architectural framework, has been pushed for a long time by the industry as an architecture to support multimedia services over the IP. Thus, offering VoLTE over the common IMS infrastructure that serves both wireless and wired customers of the same operator makes much sense. However, as of this writing, the IMS has not been adopted widely by LTE operators.

As an extension of VoLTE, **Video over LTE (ViLTE)** has been proposed to facilitate the offerings of face to face, interactive video services in LTE.

To satisfy the requirements and needs of the users of critical communications systems, LTE has been enhanced with additional capabilities such as D2D, ProSe, and Group Call System Enablers (GCSE). These and more of these critical communications related capabilities are discussed in Section 8.11 in more detail.

By using LTE and its capabilities to transport information (voice, data, video, etc.), a countless number of applications can be offered by carriers and third parties.

8.5 OPERATIONS

By design, LTE is a **trunked network**. Technically, nothing prevents an organization from deploying an LTE network in a conventional architecture. However, this mode of operation would be overkill. It may be somewhat costly and not efficient.

An unusual mode of operation, called **LTE-D**, was developed for LTE as part of Release 12. This is especially important for critical communications use. LTE-D allows so-called D2D communications without the eNBs and EPC getting involved. This results in efficient use of the spectrum and the energy, and faster peer-to-peer location-specific applications and services since there is no need for system level or GPS-based apps for location-based services. Up to 1000 devices in about 500 meters proximity can be engaged in the discovery process for potential D2D communications. The only restriction is that line-of-sight is required for D2D communications.

LTE operations are also discussed under the following two categories: FDD and TDD. LTE is capable of using all of these modes.

LTE-FDD uses so-called "paired spectrum." Two symmetric, separated, frequencies are used—one for the uplink and the other for downlink communications channels. If we assume that each link is 10 MHz wide, then the radio frame carrying information is used on both links simultaneously.

This mode of operation is more suitable for those applications where data rates on uplink and downlink channels are symmetrical.

This mode is commonly used in previous generations (2G, 3G) of cellular networks and LTE-FDD is designed to be a transitional mode as the carriers move from 2G and 3G services to LTE-based services.

LTE-TDD, also known as **TD-LTE** in some countries, uses so-called "unpaired spectrum." Both uplink and downlink channels use the same frequency. However, they use different time slots for transmitting their data. In this mode, both uplink and downlink channels use the entire bandwidth (say 20 MHz). The radio frame is divided into time slots (say 0–9) and each slot is assigned either for uplink or downlink. Therefore, the ratio between the capacities of downlink and uplink channels can be changed dynamically. Compared to LTE-FDD, the interference in LTE-TDD may become an issue since both the transceiver and the receiver operate on the same frequency, which results in higher spectrum efficiency.

This mode of operation is more suitable for those applications where data rates on uplink and downlink channels are asymmetrical.

8.6 SECURITY

Security in LTE is considered much better than that in many other wireless communications systems. LTE components and protocols embed a rich set of security features and procedures. However, since a majority of these features are optional, many LTE products may not have these optional features implemented. Furthermore, some of the security threats are not addressed at all. Some of these threats may have a significant impact on critical communications systems that are based on LTE technology. Reference [14] addresses LTE security within the context of public safety networks and identifies the following threats:

- General computer hardware and software-related security threats—The components of the LTE networks are mostly off-the-shelf-commercial products, which are considered less secure.
- Renegotiation attacks—The users may be forced by an illegally placed (or attacked) base station to downgrade to an older technology, which may not have the same security features. As a result, calls can be intercepted easily.
- Device and identity tracking—These identities can be obtained by using illegal methods. End-user devices and subscribers can be tracked by using these stolen IDs.
- Jamming UE radio interface—This threat makes radio transmission impossible due to the interference caused by jamming.
- Attacks against the secret key (K)—The stolen key (illegally obtained from the network components or elsewhere) may be used to access the network and its services illegally.
- Availability attacks on eNodeB and core—These attacks may overload the network components, thus preventing them from operating correctly (aka denial of service attacks).

Some of these threats have been addressed in LTE standards such as TS 33.401 (see Appendix C for a complete list of these standards). However, some threats such as jamming are not addressed. Fortunately, LTE is an evolving technology and stakeholders have been working on these issues.

Security functions for both signaling and user data are the responsibility of the MME. When a UE attaches with the network, mutual authentication of the UE and the network is performed between the UE and the MME/HSS. This authentication function also establishes the security keys that are used for encryption of the bearers.

LTE uses an Authentication and Key Agreement (AKA) protocol to allow user devices to access the network. Only after the AKA protocol is completed, encryption keys are generated to encrypt the transmissions of signaling and user data.

Air interface between the UE and eNodeB is protected by using some encryption procedures, each of which is based on the national/regional encryption algorithm. For example, in the USA, the Advanced Encryption Standard (AES) is used.

LTE also provides protection for the backhaul network (i.e. S1 interface). Security Gateways (SEGs) and IPSEC tunneling between SEG and eNodeB are two essential hardware-based protection features for the backhaul section of the LTE network.

8.7   SPECTRUM

LTE may be used in a variety of frequency spectrum bands. The process of defining and using these bands is as follows: LTE standards define, from a technical point of view, a wide range of spectrum bands (e.g. LTE Band 1 and LTE Band 4) for use in LTE-based services. However, ITU-R at its meetings, called World Radio Conferences (WRCs), makes the actual international allocations to establish some global understanding on the allocation and use of these bands. It has been proven that it is almost impossible to establish global bands. Therefore, the administrator of each country/region allocates the most appropriate bands for their country/region based on their regulatory policies. Some examples:

- North America (the USA and Canada): 700 MHz, 750 MHz, 800 MHz, 850 MHz, 1.9 GHz, 1.7 GHz, 2.1 GHz, 2.3 GHz, 2.5 GHz, and 2.6 GHz
- South America: 2.5 GHz
- Europe: 700 MHz, 800 MHz, 900 MHz, 1.8 GHz, and 2.6 GHz
- Asia: 800 MHz, 1.8 GHz, and 2.6 GHz
- Australia and New Zealand: 1.8 GHz and 2.3 GHz

LTE also permits the allocation of these bands to two different users—a *primary user* who has the first right to use and a *secondary user* who can use the band when the primary user is not using it.

As discussed in Section 8.5, LTE-FDD requires two symmetrical frequency bands (one for uplink and another one for downlink for each user), whereas LTE-TDD requires only one band for both uplink and downlink. Therefore, allocation of LTE frequency bands differs depending on the type of operations. For example, LTE Band 1 through LTE Band 22 are allocated for LTE-FDD operations; and LTE Band 33 through LTE Band 41 are allocated for LTE-TDD operations.

The ever-increasing demand for mobile communications always puts additional pressure on the regulatory policymakers and international standards developing organizations. As a result, additional frequency bands have been added regularly, albeit new bandwidths are small (around 10–20 MHz) compared to the original ones. However, through a new LTE-Advanced feature called channel aggregation, some of these small bandwidths in different frequency bands may be combined to achieve communications with a much higher total aggregated bandwidth, say 100 MHz.

8.8 STANDARDIZATION

The LTE specifications have been developed and standardized by 3GPP, a collaborative organization established by a group of telecommunications associations (initially, it was set up to specify 3G mobile phone systems, but later on, its scope was expanded to include IMS, LTE, and LTE-Advanced). The LTE initiative was established by 3GPP in 2004 and standardized within the Release 8 feature set. The first commercial LTE system was deployed in late 2009. 3GPP working groups added new features and technology components into later releases to improve LTE. Table 8.2 below shows the versions of releases and the release dates.

Release 9—The primary focus of this package is on LTE/EPC enhancements. In particular, dual-layer beamforming, self-organizing network (SON), enhanced HeNB (LTE femtocells), positioning, emergency services, and evolved Multimedia Broadcast Multicast Services (eMBMS) were added to the standard.

Release 10—LTE-A was introduced in this release to fulfill the requirements set by the ITU for IMT Advanced, also referred to as 4G. This package included enhanced peak data rates (1 Gbps downlink/500 Mbps uplink) and increased spectrum efficiency (up to 100 MHz). Other features of Release 10 include carrier aggregation, advanced MIMO (up to MIMO 8 × 8 in downlink and up to MIMO 4 × 4 in uplink), and enhanced Intercell Interference Coordination (eICIC).

Release 11—This release includes some refinements to the features introduced in Release 10, such as additional carrier type to enhance spectrum efficiency, multiple

TABLE 8.2. LTE Standards [1, 12]

Version	Released	Info
Release 8	March 2009	First LTE release
Release 9	March 2010	LTE Enhancements
Release 10	Sept. 2011	LTE Advanced
Release 11	March 2013	LTE-A Enhancements
Release 12	March 2015	LTE-A Enhancements
Release 13	March 2016	Critical communication+other enhancement
Release 14	June 2017	LTE-A Pro

timing advances, and Further eICIC (FeICIC) for devices with interference cancellation. Release 11 also introduced Coordinated Multipoint transmission and reception (CoMP) to improve cell-edge user data rate and spectral efficiency.

Release 12—This package contains several significant enhancements to LTE-A. The following is a list of several primary features of Release 12:

- Enhanced small cells for LTE
- Several new features to improve the support of Heterogeneous Network (HetNet) mobility
- D2D communication
- New Carrier Type (NCT)
- Further enhancements to TDD for downlink/uplink Interference Management and Traffic Adaptation (eIMTA)
- Further downlink MIMO enhancement
- Coverage enhancement
- Enhanced inter-working solutions between LTE and Wi-Fi

Release 13—This release focuses on the enhancements to LTE-A as well as new features specific to critical communications applications. This package contains features like Machine to Machine (M2M) communication, Mission Critical Push to Talk over LTE (MC-PTToLTE), the isolated E-UTRAN operation for public safety, and enhanced LTE carrier aggregation.

Release 14—This release includes the features marked as "LTE-Advanced Pro" features, which include [16]:

- Energy efficiency
- Location Services (LCS)
- Mission-critical data over LTE
- Mission-critical video over LTE
- Flexible Mobile Service Steering (FMSS)
- Multimedia Broadcast Supplement for Public Warning System (MBSP)
- Enhancements for TV service
- Massive Internet of Things (IoT)
- Cell Broadcast Service (CBS)

As far as the standardization of critical communications related features is concerned, 3GPP has been collaborating with various organizations and the vendor community within the public safety sector. As a result of this collaboration, 3GPP created a group, named the System Architecture 6 (SA6) group, to focus on application layer functions specific to critical communications. Many of the features specified by this group were incorporated into Release 13 and Release 14. Some key features in this work include:

- High-Power User Equipment (HPUE)
- MC-PTT
- QoS class identifier
- Enhancements to proximity-based services
- eMBMS
- Isolated E-UTRAN operations
- Mission-critical data and video requirements [15]

Standardization work related to critical communications applications in 3GPP is expected to continue and to be included in future releases.

8.9 FUTURE

LTE has proliferated and is projected to increase to around 1.3 billion users by 2018. Mobile broadband services are now centered on LTE. However, data traffic on wireless networks has been doubling annually, and the overall data traffic is expected to grow 12-fold by the end of 2018 [16]. The current LTE technology will not be capable of carrying this rapid increase of data consumption. To address these challenges, 3GPP and many other standards developing organizations have been busy in specifying the next generation of wireless mobile networks by not only enhancing the existing LTE features but also by adding many new advanced features.

To many, the answer is 5G, which is a catch-all term used to refer to numerous improvements on LTE and all kinds of brand new advanced features. 5G is envisioned to increase the capacity and performance of LTE-based networks by several orders of magnitude compared to the current systems [17]. 5G and other technologies that may play critical roles in future critical communications are discussed further in Chapter 9.

8.10 DEPLOYMENT

LTE as a commercial broadband wireless communications technology has been deployed widely all around the world. Reference [18] reports that there are 1.29 billion LTE subscribers worldwide. It also reports that as of June 2016, there are 118 LTE-Advanced networks in 54 countries.

An LTE network investment requires a detailed deployment strategy based on some critical factors including spectrum availability, the architectural approach, coverage, capacity, and available device ecosystem. Several significant cost factors should be considered seriously. Some of these cost factors are related to the initial

TABLE 8.3. Cost Factors Impacting LTE

CAPEX	OPEX
RAN and core network planning and design	Upgrades of hardware and software
Mobile backhaul design	Operation and maintenance
Site acquisition and deployment	Training
Hardware and Software	Headcount
OSS tools and applications	Management
Installation and integration	
Verification and testing	
Training and facilities	

deployment, and some are related to the ongoing operational cost. These factors are summarized in Table 8.3.

The most significant cost driver of LTE deployment is the radio access network, which can represent up to 70% of the total Capital Expenses (CAPEX). The site acquisition and site deployment-related costs are about 60% of the total cost associated with the radio access network. An efficient site deployment strategy should also consider maximizing coverage with less number of sites. This will help to reduce CAPEX significantly.

As discussed above, the cost of building a new LTE-based wireless communications network is significant. Many billions of dollars will be needed. Therefore, the use of the existing infrastructure at least for some of the network's components is highly crucial.

8.11 USE OF LTE AS A CRITICAL COMMUNICATIONS NETWORK

Release 12 will enhance LTE to meet public safety application requirements [19]. Two critical public safety-related study items, direct mode operation and group call functions, are included in Release 12 [11]. The primary objective of direct mode operation is to facilitate the ProSe that provide access to network services outside the usual network coverage area. In this case, one mobile acts as a relay point between two others and allows communication to take place without going via the network [20–22]. LTE group call system enablers will optimize and support dynamic groups with mobile users and dispatchers often operating in a push to talk mode.

Public safety agencies and other stakeholders (such as the TETRA Critical Communications Association [TCCA], the Association of Public Safety Communications [APCO], and First Responder Network Authority [FirstNet]) around the world are working to drive the development of additional features on top of the core LTE functionality that is typically associated with public safety systems [17]. In response,

3GPP has been enhancing existing features and adding new features directly related to critical communications applications, as evidenced in Release 13 and Release 14. Some primary examples taken from [15] are:

- HPUE—To increase radio coverage, which is desperately needed for rural areas to provide more affordable critical communication services, there is a need to increase the transmission power. The HPUE features increase the transmission power to 1.2 watts (from 200 milliwatts).
- MC-PTT—This is a "must-have" feature for critical communication applications. This feature requires the one-to-many communications capability, which commercial wireless systems do not provide. Addition of this feature to LTE releases has been ongoing. Several significant changes to LTE have been specified. These changes include provisioning, identity management, group management, security, etc.
- QoS class identifier (QCI)—This is another essential capability required for commercial LTE networks providing services to public safety agencies. This capability allows the network to be able to differentiate the traffic specific to public safety communications from other types of traffic. By assigning the highest priority class ID to public safety communications, the operators and service providers can satisfy the QoS values required for the public safety traffic. This capability has a significant impact in implementing some other critical communications capabilities (such as MC-PTT) in LTE.
- Enhancements to ProSe—First responders would like to have device-to-device communication to carry out their operations, especially in large-scale emergency situations where the communications infrastructure is down. D2D, direct mode, and LTE-D protocol, under the umbrella term "ProSe" have been included in LTE standards. LTE-D especially can provide the availability status of the other users around a user device. This means that the user is now able to establish communication with those available users nearby. Note that these features are also discussed in Section 8.5.
- GCSE—It is a capability to help improve the use of eMBMS.

The following is a list of additional LTE features that are directly related to the operations of critical communications users:

- Mission-critical data and video
- Isolated E-UTRAN operations
- Fast call setup
- Emergency call
- Group management and fleet management
- Late entry (to a group call already in progress)
- Area selection—a group call based on the location of individual subscribers

- Call authorized by dispatcher—the ability of a central dispatcher to control who can talk and when (control room functionality)
- Encryption
- Enhance the resilience of LTE networks for public safety applications

Note that most of these features do exist in some form and they are continuously enhanced in each subsequent 3GPP release.

LTE as a broadband technology for critical communications systems has gained wide acceptance. The following are examples of LTE networks being deployed for public safety:

USA—As of this writing, the planning and deployment of a nationwide public safety mobile broadband network are underway. In February 2012, the US Congress passed the "Spectrum Act" legislation [23, 24] to build a nationwide public safety mobile broadband network. This act resulted in the creation of FirstNet, an independent agency, established within the National Telecommunications and Information Administration (NTIA) [25, 26]. This legislation includes critical changes to public safety mobile broadband communications such as dedicated spectrum, 20 MHz in the 700 MHz spectrum range for public safety broadband use, a single foundation for nationwide emergency and daily public safety communications, and a nationwide private mobile broadband network based on LTE.

China—China has been trialing an LTE solution in the 1.4 GHz band for public safety by using Nokia's TD-LTE technology (TD-LTE is the same as TDD-LTE; both TD and TDD refer to "time division duplex"). There have been other LTE-based public safety activities involving the Guilin police and the Nanjing government as well.

England—The UK Home Office has been actively working on plans and deployment for an LTE-based public safety network that could replace the existing TETRA-based systems in Great Britain.

Qatar—The State of Qatar Ministry of Interior was one of the first public safety entities to commercially award a public safety LTE network contract in the Middle East. It is now fully operational and providing data services to security departments. The network is being used for fixed and mobile video applications [16].

Also, there are many other LTE-based public safety communications systems related activities in other countries as well, such as those by the following agencies:

- European TETRA association
- Sao Paulo Military Police Force and Brazilian Army for World Cup

- Hong Kong Police
- New Zealand/Australian protection forces
- Abu Dhabi Police
- Canadian Interoperability Technology Interest Group (CITIG)
- Dutch Police
- Venezuela Police

Public safety broadband networks are rapidly gaining momentum, and almost every equipment vendor is involved to a certain degree [27, 28].

8.12 SUMMARY AND CONCLUSIONS

This chapter provided a detailed discussion on LTE technologies, including LTE-A and LTE-A Pro. Since LTE has been covered in great detail in the literature, this chapter focused on the aspects related to critical communications systems. The use of LTE as the broadband technology for public safety networks has been considered for several years in many countries.

LTE is a widely accepted technology for mobile broadband communications today. LTE offers several important benefits compared to 3G, including reduced latency, higher user data rates, improved system capacity, and better coverage. In addition to the improvements in radio access technologies, packet core networks have also evolved into a flat, all IP-based architecture.

An LTE network consists of an access network called the E-UTRAN, and the core network called the EPC. The E-UTRAN consists of base stations (eNBs), which are connected to the user devices wirelessly through a standardized air interface. The EPC consists of some nodes, which are connected to each other through standardized wireline interfaces.

To address the needs of critical communications users, LTE has been enhanced with additional capabilities such as D2D, ProSe, and GCSE. Some other key features in this work include:

- HPUE
- MC-PTT
- QoS class identifier
- Enhancements to proximity-based services
- eMBMS
- Isolated E-UTRAN operations
- Mission-critical data and video requirements

LTE specifications have been developed and standardized by 3GPP within the Release 8 feature set. 3GPP has been enhancing the existing features and adding new features in subsequent releases. LTE-A was the marker used with Release 10. LTE-A Pro was another marker being used with Release 14.

LTE as a commercial broadband wireless communications technology has been deployed widely all around the world. LTE as a broadband technology for critical communications systems has gained wide acceptance as well. An example is the planning and deployment of a nationwide public safety mobile broadband network called FirstNet in the USA.

LTE has grown rapidly and is used almost ubiquitously around the world to access the Internet. As a consequence, the current LTE technology will not be capable of carrying this rapid increase of data consumption. To address these challenges, it is necessary for LTE to evolve further into 5G to increase the capacity and performance of LTE-based networks by an order of magnitude compared to the current systems.

REFERENCES

1. 3rd Generation Partnership Project, "Mobile broadband standard: LTE." [Online]. Available: http://www.3gpp.org/technologies/keywords-acronyms/98-lte. [Accessed: Jul. 24, 2016].
2. E. Dahlman, S. Parkvall, and J. Skold, *4G LTE/LTE-Advanced for Mobile Broadband*. Academic Press of Elsevier, 2011.
3. S. Palat and P. Godin, "LTE network architecture: A comprehensive tutorial," Alcatel Lucent, Strategic White Paper, 2009.
4. TETRA and Critical Communications Association, "TETRA and LTE working together, v1.1." White Paper, Jun. 2014. Available: https://tcca.info/fm_file/tetra-and-lte-working-together-v1-1-pdf/.
5. Motorola Solutions, "The future is now: Public safety LTE communications," White Paper, Aug. 2012.
6. C. Cox, *An Introduction to LTE: LTE, LTE-Advanced, SAE, VoLTE and 4G Mobile Communications*, 2nd ed. John Wiley & Sons, 2014.
7. C. Gessner, *Long Term Evolution: A Concise Introduction to LTE and its Measurement Requirements*. Rohde & Schwarz Publication, 2011.
8. M. Poikselkä, H. Holma, J. Hongisto, J. Kallio, and A. Toskala., *Voice Over LTE (VoLTE)*. John Wiley & Sons, 2012.
9. Qualcomm, "The rise of LTE-Advanced Pro," White Paper, Feb. 2017.
10. M. Rowe. (2017, May 3). LTE-Advanced Pro: The bridge to 5G. *EDN* [Online]. Available: http://www.edn.com/electronics-blogs/rowe-s-and-columns/4458335/LTE-Advanced-Pro--The-bridge-to-5G. [Accessed: Sep. 26, 2017].
11. W. Lehr and N. Jesuale, "Spectrum pooling for next generation public safety radio systems," in *IEEE Proc. Dynamic Spectr. Access Netw. (DySPAN2008) Conf.*, Chicago, Oct. 2008, pp. 1–23.

12. "LTE ue-Category." [Online]. Available: http://www.3gpp.org/keywords-acronyms/1612-ue-category. [Accessed: Sep. 2, 2017].
13. J.-G. Remy and C. Letamendia, *LTE Services*. Wiley-ISTE, 2014.
14. J. Cichonski, J. M. Franklin, and M. Bartock, "Guide to LTE security," National Institute of Standards and Technology, Gaithersburg, MD, NIST SP 800-187, Dec. 2017.
15. E. Olbrich, "The timeline for public-safety LTE standards elearning," Radio Resource Magazine, 2018. [Online]. Available: https://www.rrmediagroup.com/eLearning/frmSignin/MCID/167. [Accessed: Sep. 15, 2018].
16. Signals and Systems Telecom, "The public safety LTE & mobile broadband market: 2012–2016," Market Report, Nov. 2012.
17. 3rd Generation Partnership Project, "Delivering public safety communications with LTE," Jul. 2013. [Online]. Available: http://www.3gpp.org/news-events/3gpp-news/1455-Public-Safety. [Accessed: Sep. 15, 2018].
18. "List of 3G/4G deployments worldwide (HSPA, HSPA+, LTE." [Online]. Available: http://www.4gamericas.org/en/. [Accessed: Jul. 24, 2016].
19. 3rd Generation Partnership Project, "3GPP release 12 LTE." [Online]. Available: http://www.3gpp.org/specifications/releases/68-release-12. [Accessed: Sep. 15, 2018].
20. TCCA, "TCCA liaison to 3GPP SA on group communications and proximity services," 3GPP SP-120456, Jul. 2012.
21. K. Doppler, M. Rinne, C. Wijting, C. Ribeiro, and K. Hugl, "Device-to-device communication as an underlay to LTE-Advanced networks," *IEEE Commun. Mag.*, vol. 47, no. 12, pp. 42–49, Dec. 2009.
22. P. Janis, C.-H. Yu, K. Doppler, C. Ribeiro, C. Wijting, K. Hugl, O. Tirkkonen, V. Koivunen, "Device-to-device communication underlying cellular communications systems," *Int. J. Commun. Netw. Syst. Sci.*, vol. 2, no. 3, pp. 169–178, 2009.
23. "Spectrum management for the 21st century: The President's spectrum policy initiative," National Telecommunications Information Agency, Mar. 2008.
24. Federal Communications Commission, "FCC takes action to advance nation-wide broadband communications for America's first responders," News Release, Jan. 2011.
25. First Responder Network Authority, National Telecommunications and Information Administration, "FirstNet." [Online]. Available: https://www.firstnet.gov/. [Accessed: Sep. 15, 2018].
26. National Public Safety Telecommunications Council, "Defining public safety grade systems and facilities—final report," NPSTC Public Safety Communications Report, May 2014.
27. A. M. Seybold, "Seybold's take: Public safety's 700 MHz LTE network an opportunity for vendors," Fierce Wireless, Mar. 14, 2012. [Online]. Available: https://www.fiercewireless.com/wireless/seybold-s-take-public-safety-s-700-mhz-lte-network-opportunity-for-vendors.
28. Alcatel-Lucent, "Alcatel-Lucent and first responders conduct trial of 4G LTE public safety broadband mobile network," Alcatel-Lucent Press Release, Nov. 2013.

9

FUTURE TECHNOLOGIES FOR CRITICAL COMMUNICATIONS SYSTEMS

This chapter provides a brief introduction to some of the emerging technologies that public safety and mission-critical applications may use. Although there are many new emerging technologies, in this chapter, we focus on fifth generation (5G) wireless mobile communications systems, Augmented Reality (AR), the Internet of Things (IoT), and Big Data analytics, which are expected to make a significant impact on public safety and critical business applications.

9.1 INTRODUCTION

There is an arsenal of emerging technologies such as 5G, AR, Artificial Intelligence (AI), and the IoT that are expected to have significant impacts on several application areas. These technologies may also play a significant role in improving the effectiveness of public safety and mission-critical personnel, especially first responders. Also, the new technologies may drastically change the way public safety activities are handled.

Fundamentals of Public Safety Networks and Critical Communications Systems: Technologies, Deployment, and Management,
First Edition. Mehmet Ulema.
© 2019 by The Institute of Electrical and Electronics Engineers, Inc. Published 2019 by John Wiley & Sons, Inc.

5G, 5th generation wireless mobile communication systems, is being defined as the successor of Long Term Evolution (LTE) to be able to address the ever-increasing demand for much higher data rates and much lower latency, among other requirements. Additionally, other technologies such as the IoT, AR, edge computing, Software-Defined Networking (SDN), and Network Functions Virtualization (NFV) may become part of 5G and commercially available.

The use of new mathematical and statistical tools such as machine learning and deep learning to deal with so-called "Big Data" will bring entirely new approaches to handling public safety activities. Coupled with AI, cloud computing, and edge computing, these new predictive analytics tools may change the way criminal investigations are carried out. Social media, the Internet, and other data sources may be used by these new predictive crime analytics tools to make much more accurate predictions in law enforcement activities.

Other emerging technologies such as thermal imaging, Unmanned Aerial Vehicles (UAVs) (aka drones), robotics, and smart devices (e.g. smart guns) may be used by first responders as well. There are already smart gun products on the market [1]. Some of these guns include sensors, communications links (LTE, Bluetooth, GPS, etc.), biometric fingerprint sensors, and Radio-Frequency Identification (RFID) chips to provide data (time, place, the person using the gun, etc.) recording, location tracking, and authorization for use.

Smart buildings, smart spaces, and smart cities are other emerging technologies that could be used by public safety personnel. For example, an appropriately equipped (sensors, cameras, etc.) campus can provide the first responder the exact location of the incident (gunshot, terrorist activities, etc.) that requires an emergency response.

9.2　5G AND BEYOND

Every decade or so a new generation of technologies has been introduced to take wireless mobile communications services to the next, higher level. The current generation, as discussed in Chapter 8, is called the 4th generation, aka 4G; based on LTE, it was introduced around 2010. As of this writing, the stakeholders in this sector have been occupied in making the next generation, called 5G, a reality by 2020.

So, the term 5G is a code name used to refer to technologies that will be used in the mobile wireless networks beyond LTE-based 4G networks. As expected, wish lists are too many and too long! Academic and industry researchers as well as practitioners, together with Standards Development Organizations (SDOs) have been swamped with all sort of research, development, and standardization activities. The last couple of years, we have witnessed a tug of war between the ones who have and the one who would like to have, between the visionaries and the current market leaders.

It is most likely that the resulting 5G technologies will be primarily based on LTE as an extension of LTE-Advanced (LTE-A) and LTE-A Pro. Also expected is a rather heterogonous network including small cells, IEEE 802.11ac (and its successors) integration (seamless roaming between Wireless Fidelity [Wi-Fi] and macro cells), with IoT capabilities. Higher frequency spectrum beyond 6 GHz is expected to be included later on. Higher frequencies are needed to be able to provide more capacities at higher data rates. Research and experiments at millimeter (mm) wave spectrum along with massive Multiple-Input Multiple-Output (MIMO) and beam steering can be achieved with phased array technology, using a large number of antenna elements to show that higher data rates and lower latencies are possible at frequencies much above 6 GHz rate [2].

5G New Radio, or NR, is the new standard being developed by the 3rd Generation Partnership Project (3GPP) for 5G wireless technology, capable of much faster, efficient, and scalable network. 5G NR will be able to improve capacity, density, spectrum, and network efficiency, and will allow devices to support millimeter waves [3].

The infrastructure is expected to incorporate SDN and NFV and energy efficient information and communications technologies. Figure 9.1 highlights the areas of 5G infrastructure being discussed [4].

The following are key challenges and requirements for 5G infrastructure [4]:

- Supporting network speeds at around 20 Gbps or higher, enabling a user to download, for example, a full High Definition (HD) movie in seconds, and streaming high resolution (8K) videos
- Having a latency in millisecond levels; this is essential for future technologies like AR, driverless vehicles, and tactile Internet to work effectively

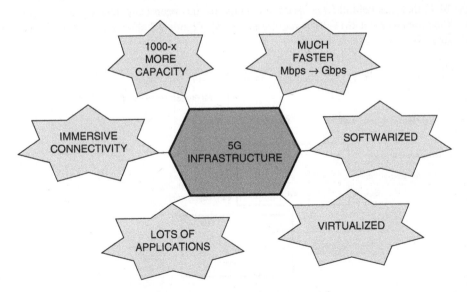

Figure 9.1. 5G infrastructure expectations.

- Providing 1000 times higher wireless area capacity and more different service capabilities compared to 2010
- Saving up to 90% of energy per service provided; the primary focus will be on mobile communication networks where the dominating energy consumption comes from the radio access network
- Reducing the average service creation time cycle from 90 hours to 90 minutes
- Creating a secure, reliable, and dependable Internet with a "zero perceived" downtime for services
- Facilitating very dense deployments of wireless communication links to connect over 7 trillion wireless devices serving over 7 billion people
- Ensuring for everyone everywhere access to a broader panel of services and applications at a lower cost
- Ensuring dense connectivity for machines, automobiles, city infrastructure, public safety, and more

Enhancements to both system management and Operational Support Systems (OSSs) are also expected benefits of 5G. They are both supposed to improve reliability, availability, serviceability, resilience, consistency, analytics capabilities, and operational efficiency of these systems, which will reduce Operational Expenses (OPEX).

According to the International Telecommunications Union (ITU), a set of criteria, called International Mobile Telecommunications (IMT)-2020, will be developed by the end of 2017 and SDOs will be invited to submit their candidate proposals to the ITU to be tested against IMT-2020 requirements. Meanwhile, 3GPP, the leading SDO that was behind LTE-based 4G standards, has been busy developing 5G specifications to meet IMT-2020 requirements [5]. Figure 9.2 shows a timeline for this process.

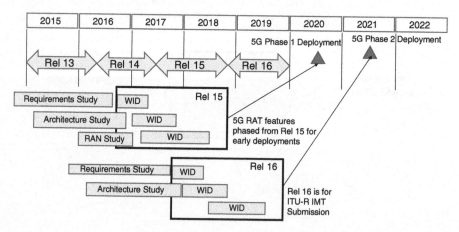

Figure 9.2. 3GPP roadmap for 5G [5].

5G technologies are expected to be the facilitator and enabler of many vertical and professional applications, including automobiles, entertainment, the medical field, finance, and public safety. As far as public safety applications are concerned, the following 5G capabilities are of more vital importance: ultra-reliability and security, low latency, user prioritization, sensors and the Internet of Things, and radio performance [6]. In general, it is expected that 5G technologies will enable robust, mission-critical, interoperable public safety communications, allow effective and efficient spectrum use and sharing, enable advanced communications technologies, and help public agencies improve their response to emergencies.

One of the primary use cases that was discussed for 5G by the stakeholders is public safety communications and applications. Therefore, 3GPP plans to include essential functionalities that are expected to be utilized by future critical communications systems empowered by 5G technologies. Some of the capabilities considered along these lines include body camera support, instant data access, push-to-talk, push-to-video, haptic computing features such as body part movement based actions, loudspeaker phone, long battery life, and hardened devices operating in harsh environments [7].

5G may help to improve traffic safety by enabling road safety services including Vehicle-to-Vehicle (V2V) and Vehicle-to-Everything (V2X) communications and automatic and wireless control of emergency stops to prevent accidents. 5G may accelerate the trend to move towards smart and autonomous devices.

With relatively much higher data rates and much lower latency, 5G, together with augmented and virtual reality and IoT capabilities, will facilitate public safety agencies' access in real time to significant amount of field data including ultra-high resolution video, providing greater on-the-scene detail to help the agencies to take better and timely actions. Reference [8] shows that a 1-minute improvement in response time translates to a reduction of 8% in mortality, and reduction in crime and the cost of law enforcement.

There is no question that 5G-based critical communications systems will form a formidable infrastructure for public safety agencies. However, this infrastructure needs to be augmented with a variety of capabilities, applications, and services that will help the agencies to be much more effective especially in handling emergency situations in disaster recovery operations. As of this writing, this is an open and attractive research area. The following is a list of some of the areas to consider in 5G and beyond:

- Relaying and Device to Device (D2D) communication in 5G for emergency/disaster management, especially important when the infrastructure is down
- 5G-based UAVs, drones, robots, and networked sensors for emergency management, especially important when the disaster areas are hard to reach for rescue personnel
- Energy efficiency, energy harvesting, and the use of other energy sources in critical communication networks when the power is down during the disaster

9.3 AUGMENTED REALITY (AR)

To understand AR, it is useful if we discuss Virtual Reality (VR) first. VR technology and its products (helmets, goggles, etc.) are used to allow the user to be placed in a virtual environment, which is mainly designed/created for a specific application. The user is isolated from the real world and just interacts with various elements (animals, human-like symbols, etc.) in an imaginary, fabricated world.

On the other hand, AR combines virtual reality with the real environment. The user of an AR product is always in touch with the real world, while interacting with virtual, created contents in the real world that we see and live in, so the experience is real and contextual.

So, we can say that AR is "the interaction of superimposed data, graphics, audio and other sensory enhancements over a real-world environment that is displayed in real time [9]."

AR devices include smartphones and tablets, other mobile devices specially designed for a specific workforce, Head-Up Displays (HUDs) (for windshields, visors, etc.), Head-Mounted Displays (HMDs), eyeglasses, goggles, helmets, contact lenses, and virtual retinal displays. A person wearing an AR device while walking may interact (by waving hands) with information tagged to objects (people, things, etc.) [10, 11].

AR combined with smart cities, the IoT, and Big Data will be very beneficial and useful, essentially becoming a visual portal for many private and public applications, including public safety and emergency services.

The following is a list of potential areas in which firefighters, first responders, and other critical communications personnel may use AR to be more effective in emergency situations:

- AR head-up windshield displays on vehicles providing route guidance and real-time sensor data on environmental and hazardous conditions
- Helmet-mounted AR devices and visors allowing first responders to see and hear through smoke, fire, rubble, poor weather, and other conditions
- AR disaster applications providing visual and audio guidance to citizens seeking refuge, evacuation routes, or emergency assistance in a disaster situation
- Real-time data-driven AR applications allowing law enforcement officers to access location-specific information and data on dangerous situations via smart glasses, in-vehicle displays, and other wearables
- Pointing AR-equipped mobile devices at a building, down a street, or over an entire community, personnel can access authorized geo-specific data on crime statistics and other environmental factors. By just pointing to a door, information about the people who pass through it and recorded videos can be viewed instantly.

AR can be a useful, risk-free tool for training critical communications personnel, especially firefighters and first responders [12]. Trainees with hands-free, head-up AR glasses can see live video and data in a simulated environment. This provides much better situational awareness, not only for the trainees but also for the instructors. With these kinds of AR tools, activities of various response teams can be monitored, remotely and locally, with more accurate information so that much more efficient coordination can be established. Furthermore, the instructors can intervene in real time by providing guidance and introducing new obstacles into the simulated scenarios.

As of this writing, there are some AR devices on the market, such as Google Glass. They are mostly early versions or beta versions, but there is significant expectation that the AR market will grow to be worth 61 billion US Dollars (USD) by 2023 [13]. Several technical and economic obstacles need to be overcome before AR devices become useful for critical communications applications. Higher cost, heaviness, limited range, limited battery life, software unavailability, and limited field view are some of the problems associated with these early versions of AR devices on the market.

9.4 INTERNET OF THINGS (IoT)

The IoT has become a favorite topic during the past several years, because the term IoT is used in many different contexts—from sensors (wireless and mobile) to smart cities. However, as the name implies, it is primarily about connecting "things" through the Internet. *Things* include physical objects like consumer devices embedded with electronics and software, smart devices, sensors, and actuators, which can move and take actions such as pushing buttons [14, 15].

Some people have included people and others in the mix and called it the Internet of Everything (IoE) [16]. So, *everything* connected through the network exchanges data, perhaps in real time, to provide services to the users.

As alluded to above, there are numerous application areas for the IoT, including connected smart homes, connected smart cities, connected smart infrastructures (roads, energy grids, public facilities, etc.), connected vehicles, connected wearables, and the industrial Internet.

In this book, naturally, we are more interested in the use of the IoT technology in critical communications applications such as locating emergencies, tracking health signs of the injured in emergency scenarios, and health tracking of first responders. The IoT can be helpful in improving safety and security, increasing response times, and increasing productivity and efficiency of critical communications agencies [16].

Sensors embedded in the body and uniforms of public safety personnel as well as cameras connected to the personnel may be used to monitor their location, health, and other pertinent information that will be useful in the overall effectiveness of their operations. Furthermore, the information, which could be in real-time and multimedia, can be shared and routed promptly among the right personnel to provide, for example, better situational awareness and faster and more intelligent decision making.

Also, sensors can be embedded in stationary or mobile devices such as robots and UAVs to support emergency situations that would be dangerous for the agents to enter and search. For example, UAVs can be used in hard to reach disaster areas to collect and disseminate real-time data to be used in rescue operations by robots and the personnel on the ground.

9.5 BIG DATA ANALYTICS

The term *Big Data* was coined to emphasize the exponentially growing nature of the data available thanks to the advances in computers and the availability of emerging technologies such as the Internet of Things (IoT) and augmented reality. The *volume* of data, along with its *velocity* and *variety* form the so-called *3Vs* that characterize Big Data [17].

Before we discuss how Big Data analytics can be helpful for public safety agencies to carry out their business, let's explain some of these terms first.

It is incredible that so much relevant information can be measured so quickly and stored in the form of raw data (text, sound, photos, videos, etc.). However, what is more important is what can be done with the data that matters. Reference [18] describe *analytics* as "the extensive use of data, statistical and quantitative analysis, explanatory and predictive models, and fact-based management to drive decision and actions."

Making decisions based on data collection and analysis is indeed nothing new; according to [19], the term "business analytics" dates back to the late 19th century. *Big Data analytics* is the same idea as analytics, but now we have much more data of many types (transaction data, mobile phone data, social media chatter, telemetry, and on and on), and they are generated all the time and available immediately. Big Data analytics, with a variety of statistical, mathematical tools and AI with data mining and machine learning algorithms and programs, help organizations to carry out *preventive* and *predictive* analyses used in their decision making processes. So, it is critically important for public safety agencies to use Big Data analytics to identify patterns and predict potential problems before they take place.

The current overwhelming interest in this subject boils down to one word— technology. In today's high-bandwidth, cloud-based, interconnected world, data can be captured and stored on massive scales. Also, advances in software and computer hardware are allowing organizations to perform analysis on this data at breathtaking speeds. Such processing speeds have made possible the creation and application of computer algorithms that help predict outcomes or learn to recognize patterns in data.

Reference [20] emphasizes how Big Data analytics can make the intelligence more efficient by combining analytical tools with human decision making, which results in more accurate and timely actionable information. A recent report predicts that this will create a huge market (11 billion USD by 2022) for the public safety sector [21]. Another report [22] foresees *"a robust and affordable public safety communications ecosystem incorporating the state-of-art in analytic technologies that are accurate, effective, dependable, customizable, secure, and usable."* This, of course, requires a *cultural* change in the way public safety agencies conduct their business, by including new skills, new policies, evidence-based new processes, new tools, faster decision making, etc.

The following are some use cases to highlight how Big Data analytics can help public safety agencies:

- **Content and context analysis**—it is well known that criminals and terrorists use the digital media (web, emails, messages, Facebook, Twitter, YouTube, etc.). They leave behind a wealth of data that can be captured, analyzed, mapped into a social graph, and used in crime-fighting operations. An example is the identification of the influential people and their followers in criminal and terrorist activities.
- **Situation analysis**—thanks to the significant advances in storage and computing and communications technologies, along with data mining and machine learning software and related application systems, it is possible to store, search, view, and visualize a broad set of cases spread around various jurisdictions nationwide. The resulting analytics will allow agencies to be able to see the trends, anomalies, patterns, connections, and relationships among the cases and suspects with higher accuracy and speed.
- **Predictive analysis**—the techniques and algorithms employed in this type of analyses will be helpful for predicting criminal behaviors and for predicting where and how terrorist activities may take place. Social media analytics, together with the analysis of various reports such as criminal records already available, may be used in appropriate predictive analytical models to make predictions with high confidence levels in relatively short time.

9.6 SUMMARY AND CONCLUSIONS

In this chapter, we had a brief glance at some of the emerging technologies and how they may drastically change the way public safety agencies conduct their business. We expect that these emerging technologies will combine their *forces* to develop entirely new types of systems, software, hardware, and applications for the public safety sector.

Some companies are already making significant progress along these lines. For example, as of this writing, a company claims that it has created a system that can recognize where gunshots are fired in crime scenes. The system relies on the hundreds of connected microphones deployed in an area (campus, town, city, etc.) [1]. Another noteworthy example are the advances in video and audio analytics. With the help of these technologies, the systems available today can provide, for example, facial and license plate recognition capabilities.

Government organizations, such as the National Institute of Standards and Technology (NIST) in the USA, academic researchers, as well as companies in this sector are busily addressing opportunities and challenges in these areas. For example, Reference [23] discusses real-time location-based services (indoor and outdoor) and identifies the challenges and opportunities for public safety applications on this topic. Similarly, Reference [22] focuses on Big Data analytics and again identifies the challenges and opportunities for the public safety sector on this topic over the next 20 years.

REFERENCES

1. T. Maddox, "Gunshot detection technology as part of smart city design," *TechRepublic*, Nov. 22, 2016. [Online]. Available: https://www.techrepublic.com/article/gunshot-detection-technology-as-part-of-smart-city-design/. [Accessed: Dec. 28, 2017].
2. S. Sun, T. S. Rappaport, R. W. Heath, A. Nix, and S. Rangan, "MIMO for millimeter-wave wireless communications: Beamforming, spatial multiplexing, or both?," *IEEE Commun. Mag.*, vol. 52, no. 12, pp. 110–121, Dec. 2014.
3. "First 5G NR specs approved." [Online]. Available: http://www.3gpp.org/news-events/3gpp-news/1929-nsa_nr_5g. [Accessed: Dec. 28, 2017].
4. "5G-PPP." [Online]. Available: https://5g-ppp.eu/. [Accessed: Jul. 24, 2016].
5. "3GPP." [Online]. Available: http://www.3gpp.org/. [Accessed: Jul. 24, 2016].
6. TCCA, "4G and 5G for public safety—technology options." [Online]. Available: https://tcca.info/fm_file/4g-and-5g-for-public-safety-pdf/. [Accessed: Sep. 14, 2018].
7. B. Bertenyi, "3GPP system standards heading into the 5G era." [Online]. Available: http://www.3gpp.org/news-events/3gpp-news/1614-sa_5g. [Accessed: Dec. 28, 2017].
8. E. T. Wilde, "Do emergency medical system response times matter for health outcomes?," *Health Econ.*, vol. 22, no. 7, pp. 790–806, Jul. 2013.
9. G. Curtin, "6 ways augmented reality can help governments see more clearly," *World Economic Forum*. [Online]. Available: https://www.weforum.org/agenda/2017/02/augmented-reality-smart-government/. [Accessed: Oct. 28, 2017].
10. J. Peddie, *Augmented Reality: Where We Will All Live*. Springer, 2017.
11. T. Jung, M. Claudia, and T. Dieck, *Augmented Reality and Virtual Reality: Empowering Human, Place and Business*. Springer, 2018.
12. US Ignite, "Application: Augmented reality tools for improved training of first responders." [Online]. Available: https://www.us-ignite.org/apps/AR-training-first-responders/. [Accessed: Oct. 28, 2017].
13. MarketsandMarkets, "Augmented reality market by offering (hardware (sensor, displays & projectors, cameras), and software), device type (head-mounted, head-up, handheld), application (enterprise, consumer, commercial, automotive) and geography—Global

forecast to 2023." [Online]. Available: http://www.reportsnreports.com/reports/434857-augmented-reality-market-by-component-sensor-display-software-display-type-head-mounted-head-up-handheld-spatial-application-aerospace-defense-consumer-commercial-and-geography-global-forecast-to-2020.html. [Accessed: Nov. 12, 2017].

14. B. Tripathy and J. Anuradha, *Internet of Things (IoT): Technologies, Applications, Challenges and Solutions*. CRC Press, 2017.

15. H. Geng, *Internet of Things and Data Analytics Handbook*. John Wiley & Sons, 2016.

16. Cisco, *Public Safety Justice and the Internet of Everything*. Cisco, 2014.

17. D. Laney, "3D data management: Controlling data volume, velocity, and variety," Gartner, Feb. 2001. [Online]. Available: https://blogs.gartner.com/doug-laney/files/2012/01/ad949-3D-Data-Management-Controlling-Data-Volume-Velocity-and-Variety.pdf.

18. T. Davenport and J. Harris, *Competing on Analytics: The Science of Winning*. Harvard Business School Press, 2007.

19. R. K. Klimberg and V. Miori, "Back in business," *ORMS Today*, vol. 27, no. 5, pp. 22–27, Oct. 2010.

20. G. Woodward, "How Big Data can change the world of public safety." [Online]. Available: https://www.sas.com/en_ca/insights/articles/big-data/local/how-big-data-changes-public-safety.html. [Accessed: Dec. 19, 2017].

21. Marketwire, "Big Data and data analytics in homeland security and public safety." [Online]. Available: http://www.marketwired.com/press-release/big-data-data-analytics-homeland-security-public-safety-is-forecast-reach-11b-2022-according-2181511.htm. [Accessed: Dec. 19, 2017].

22. E. Nunez, "Public safety analytics portfolio," *NIST*, Sep. 20, 2017. [Online]. Available: https://www.nist.gov/ctl/pscr/research-portfolios/public-safety-analytics-portfolio. [Accessed: Oct. 28, 2017].

23. R. Felts, M. Leh, D. Orr, and T. A. McElvaney, "Location-based services R&D roadmap," NIST PSCR, NIST Technical Note 1883, May 2015.

10

SYSTEMS AND APPLICATIONS USED IN CRITICAL COMMUNICATIONS

This chapter provides an overview of the systems used to support public safety activities. In this context, the word system refers to a computer-based entity that is composed of hardware and software with appropriate application programs to support the activities of public safety personnel.

10.1 INTRODUCTION

Supporting systems are typically located in appropriate centers to support the activities being managed at the center. Therefore, depending on the center that they support, they are equipped with appropriate hardware and software. For example, the systems used in emergency call centers have different capabilities than the systems used in command and control centers. Also, there are traditional servers that are deployed in public safety offices.

Fundamentals of Public Safety Networks and Critical Communications Systems: Technologies, Deployment, and Management,
First Edition. Mehmet Ulema.
© 2019 by The Institute of Electrical and Electronics Engineers, Inc. Published 2019 by John Wiley & Sons, Inc.

10.2 COMMAND AND CONTROL CENTERS

As mentioned in Chapter 1 briefly, command and control centers include a number of systems, applications, tools, policies, and people responsible for maintaining, administering, operating, and managing the whole critical communications network in a reliable, secure way. Depending on the country or region, there may be agency-wide or region-wide centers and systems with similar responsibilities. Ideally, all these systems should be connected and integrated to coordinate overall activities. Network and service management specific systems and operations are discussed in greater detail later, in Chapter 15. Other systems that may be integrated into this center include 911, incident management, and ticketing systems, which are discussed in the following sections.

10.3 EMERGENCY RESPONSE SYSTEMS

These systems provide a link between the people in distress (mainly life-threatening situations) and an appropriate agency such as the police, ambulance, or fire departments. People use a well-known number to dial in through wired or wireless phones to reach the system. In the USA, it is 911. Most other countries use different numbers [1]. For example, it is 112 in Europe and 999 in England and many other countries. Also, in some countries, there may be different numbers for the police, fire, and ambulance.

These systems are centralized in many countries, but in the USA, although the National 911 Office oversees the USA's 911 emergency response system [2], each municipal and county has its own system, deployed and maintained locally and jointly between the local government and the phone companies operating in the area.

Emergency response systems, together with the phone service provider's network, must be able to recognize the emergency number on any phone and forward the call to the nearest Public Safety Answering Point (PSAP). Trained 911 operators (also called dispatchers) at the PSAPs answer the call and notify (dispatch) immediately the appropriate agency, which in turn responds to the emergency. There are more than 6000 PSAPs in the USA [3]. The PSAPs may be equipped with the following equipment and tools:

- A computer-controlled system to activate pagers, used by dispatchers to alert the agencies quickly
- A mapping program, Computer-Aided Dispatch (CAD), to provide directions along with relevant route and road information and, if available, information about the caller
- Recording and storage equipment to record phone calls and all radio communications into and out of the center
- Backup generators and Uninterruptible Power Supplies (UPSs)

There are two crucial requirements here. First, the system must allow the emergency number to be entered on any phone regardless of payment. Second, the system must determine the location of the situation. The legacy **basic 911** system relied on the Public Switched Telephone Network (PSTN) to determine the subscriber's location and to get the information from the caller to identify the exact location of the situation.

With ubiquitous availability of cellular phones and IP phones, determining the caller's location has become more challenging. The current system, called **Enhanced 911 (E911)**, includes the capabilities to determine and display the location of the caller automatically through its databases and various algorithms [4]. In addition to the equipment discussed above, each E911 PSAP may also have a local network connecting databases, tools, and other equipment, as shown in the following points:

- Automatic Number Identification (ANI) to recognize the telephone number of the 911 caller and forward the information to the dispatcher (by the way, ANI is a general purpose capability used by phone companies to identify the caller and provide relevant information [5])
- Automatic Location Identification (ALI) to help the E911 system to determine the location of the phone number reported by the ALI system (similarly, ALI is another database used by phone companies [6])
- Master Street Address Guide (MSAG) to determine where to route your call [7]; it is a database, a master map, that shows the streets, street number ranges, and other address-related information

Figure 10.1 illustrates how the E911 System handles emergency calls.

Today, a vast majority of emergency calls around the world are made from cellular phones. In the USA, the Federal Communications Commission (FCC) has mandated a two-phased approach in determining the location of the wireless 911 caller [8].

- **Phase I** requires that the call be routed to the PSAP nearest to the tower with which the cell phone is communicating (this covers a rather large area that may extend from 6 to 30 miles in radius).
- **Phase II** requires an accuracy of 50–300 meters. To achieve this, there are some approaches, including network-based and phone-based ones, which use the Global Positioning System (GPS).

Also pretty common these days are the so-called Internet Protocol (IP) phones based on Voice over Internet Protocol (VoIP) technology, which work quite differently from traditional wireline phones. A significant difference is that IP phones have IP addresses, are portable, and work from any Internet connection, all of which provide challenges in determining the location of the callers.

Again, the FCC in the USA has mandated that all VoIP service providers must provide default physical location (recorded during the setup of the account) information to the PSAP.

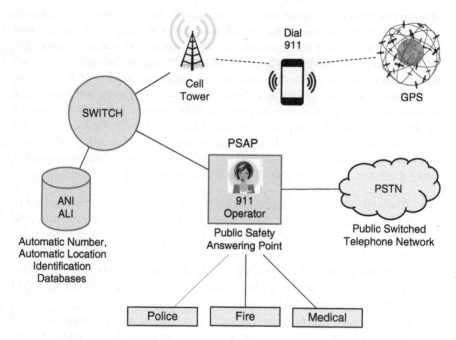

Figure 10.1. Emergency response system handling of emergency calls.

Furthermore, the US government, recognizing the impact of emerging technologies on emergency response systems, funded a new initiative, called **Next Generation 911 (NG911)**, to improve the existing E911 systems and implement a nationwide system [9].

NG911 is a faster, more resilient, much more comprehensive, and more integrated system facilitating the flow of information received from the callers to PSAPs and appropriate agencies directly and automatically. The type of the information disseminated can be in any one of or a combination of voice, photos, videos, and text messages [10].

10.4 INCIDENT MANAGEMENT SYSTEM

An incident management system provides a mechanism for all the public safety agencies in that region or country to work together "to prevent, protect against, respond to, recover from, and mitigate the effects of incidents, regardless of cause, size, location, or complexity." When all public safety agencies share the same nationwide network, interagency planning and coordination is a must, especially during incidents.

Although each agency will have its own control, command, and management centers, a unified incident control center must be established to handle the coordination and sharing of resources, capacities, etc. Again, this center will likely be equipped with appropriate systems that are used by all the public safety personnel involved in the incidents. The center should develop and provide a set of policies and procedures to describe consistently what an agency should do in response to an incident in collaboration with other agencies. The policies should include how the available capacity of the network in the incident area should be shared among the first responders.

As the name implies, incident management systems are used to deal with incidents. The primary aspects of management include detection, identification, assignment, cause, size, response, recording of all related information, analyzing, and generating a report. An incident is defined as "an occurrence, natural or manmade, that necessitates a response to protect life or property...., the word 'incident' includes planned events as well as emergencies and disasters of all kinds and sizes [11]."

In many countries, there are centralized and integrated incident management systems, which handle all types of incidents and are used by all agencies. As discussed before, in the USA, each agency in each region traditionally has its own incident management system, not necessarily integrated with the other agencies in the same region. However, there is an intense pressure to have a unified system [12] [13]. A new Department of Homeland Security (DHS) project called Unified Incident Command and Decision Support (UICDS) was launched to respond to this pressure. The resulting system, called the National Incident Management System (NIMS), is used to manage and share incident information across state and local lines, as well as with other federal agencies [11, 14]. NIMS is a set of concepts and principles that describe standardized procedures and policies for the management of all types of incidents regardless of location and complexity. NIMS has been adopted by some campuses and businesses as well.

10.5 PUBLIC WARNING SYSTEMS

Another important set of systems that are associated with public safety operations is commonly called public warning systems, which are used to inform the public about emergency situations or other matters of importance to public safety such as providing lifesaving instructions as immediately as possible. Some examples of the critical situations that the public is warned about include the following:

- Storms, hurricanes, tornados, typhoons, etc.
- Flood, landslides, tsunamis
- Wildfire, forest fires

- Air contamination, hazardous material release
- Terrorist threats, terrorist activities
- Incoming missiles
- Water contamination
- Other situations that may threaten life, property, etc.

These systems could be specialized in specific areas and therefore be used by certain specific agencies (police, fire departments, environmental agencies, etc.), or could be unified across many agencies and regions as well.

A mix of media and venues are used to send the warning. Broadcasting through television channels, radio stations, and cable networks are typical. Also used to inform the public are social media, cell phones, emails, electronic roadside displays, etc.

In the USA, there is a centralized, integrated nationwide system called the Integrated Public Alert and Warning System (IPAWS) [15]. This system is mainly controlled and managed by the Federal Emergency Management Agency (FEMA). It is the modernized alert and warning infrastructure of the USA. Various systems under regional, state, and local authorities can be integrated into IPAWS by using a standardized protocol called the Common Alerting Protocol (CAP). IPAWS provides a single interface for some other public warning systems such as the Emergency Alert System (EAS), Wireless Emergency Alerts (WEA), and the National Oceanic and Atmospheric Administration (NOAA) Weather Radio.

The WEA is especially crucial since the ubiquitous availability of wireless devices (cell phones, tablets, etc.) makes them one of the most important venues to disseminate public warnings [16]. WEA messages are broadcast directly from cell towers to all mobile devices roaming in that area. There is no need for an app. No subscription to a service is required. The message may include information such as the type and time of the alert and the action that must be taken. Figure 10.2 shows an example of a public warning message broadcast to the mobile phones in a region.

10.6 OTHERS

To support their activities, public safety agencies may use a variety of other systems. The ticketing systems used by police departments are a good example. These systems provide traditional and straightforward capabilities such as officers entering information about the traffic violation or parking violation tickets being issued, or getting a printed version of the ticket through a mobile printer. Also, some newer versions may include sophisticated algorithms and advanced capabilities—for example, detection of traffic violation events from various videos and sensors that can detect when a car passes through a red light [17].

Figure 10.2. A public warning displayed on a smartphone.

10.7 SUMMARY AND CONCLUSIONS

This chapter provided a brief discussion on some primary systems and applications used by public safety personnel. Examples include the systems used in emergency call centers, the systems used in command and control centers, incident management systems, and public warning systems.

It is expected that newer versions of these systems and applications will incorporate a variety of emerging technologies, such as high-resolution sensors, sophisticated algorithms, and big data analytics.

REFERENCES

1. US Department of State, "Emergencies—911 abroad." [Online]. Available: https://travel .state.gov/content/dam/students-abroad/pdfs/911_ABROAD.pdf. [Accessed: Sep. 14, 2018].
2. "How to become a 911 dispatcher." [Online]. Available: https://www.911dispatcheredu .org/. [Accessed: Sep. 14, 2018].

3. "Public-safety answering point," *Wikipedia*. [Online]. May 28, 2015. Available: https://en .wikipedia.org/wiki/Public_safety_answering_point.

4. "Enhanced 9-1-1," *Wikipedia*. [Online]. Nov. 26, 2017. Available: https://en.wikipedia .org/wiki/Enhanced_9-1-1.

5. "Automatic number identification," *Wikipedia*. [Online]. Sep. 9, 2017. Available: https:// en.wikipedia.org/wiki/Automatic_number_identification.

6. K. Michael, *Innovative Automatic Identification and Location-Based Services*. IGI Global, 2009.

7. R. K. Baggett, C. S. Foster, and B. K. Simpkins, *Homeland Security Technologies for the 21st Century*. ABC-CLIO, 2017.

8. L. Daniel, *Cell Phone Location Evidence for Legal Professionals*. Academic Press, 2017.

9. "About the National 911 Program." [Online]. Available: https://www.911.gov/about_ national_911program.html. [Accessed: Dec. 30, 2017].

10. "Next Generation 911 | NTIA." [Online]. Available: https://www.ntia.doc.gov/category/ next-generation-911. [Accessed: Dec. 30, 2017].

11. FEMA, "National Incident Management System; 3rd Edition," Oct. 2017. Available: https://www.fema.gov/media-library-data/1508151197225-ced8c60378c3936adb92c1a3 ee6f6564/FINAL_NIMS_2017.pdf.

12. A. M. Seybold, "FirstNet brings changes to unified incident command," *IMSA J.*, Feb. 2014.

13. J. W. Morentz, C. Doyle, L. Skelly, and N. Adam, "Unified Incident Command and Decision Support (UICDS) a Department of Homeland Security initiative in information sharing," in *Proc. IEEE Conf. Technol. Homeland Security (HST '09)*, Boston, 2009, pp. 182–187.

14. "National Incident Management System." [Online]. Available: https://www.fema.gov/ national-incident-management-system. [Accessed: Jan. 4, 2018].

15. "Integrated Public Alert & Warning System." [Online]. Available: https://www.fema.gov/ integrated-public-alert-warning-system. [Accessed: Jan. 6, 2018].

16. "Frequently asked questions: Wireless Emergency Alerts." [Online]. Available: https:// www.fema.gov/frequently-asked-questions-wireless-emergency-alerts. [Accessed: Jan. 6, 2018].

17. W. Q. Yan, *Introduction to Intelligent Surveillance: Surveillance Data Capture, Transmission, and Analytics*, 2nd ed. Springer, 2017.

11

END-USER DEVICES CONNECTED TO CRITICAL COMMUNICATIONS SYSTEMS

This chapter provides a brief discussion on the equipment used by public safety personnel. In this context, the equipment referred to are the ones linked to the critical communications network. Also, the users may be the officers on the field and dispatchers. The officers on the field may be on vehicles, boats, etc., or may be walking.

11.1 INTRODUCTION

As discussed in Chapter 1, many end-user devices heavily depend on the communications technology used in the network. For example, if the network is based on Project 25, then the user equipment must be compatible with Project 25 technology. Not only are the communications hardware and protocols different, but the device's voice, data, and video capabilities are also dependent on the limitations of the network to which they are connected. There is no question that end-user devices compatible with LTE technology have superior capabilities regarding performance, especially data and video capabilities.

Fundamentals of Public Safety Networks and Critical Communications Systems: Technologies, Deployment, and Management,
First Edition. Mehmet Ulema.
© 2019 by The Institute of Electrical and Electronics Engineers, Inc. Published 2019 by John Wiley & Sons, Inc.

The following sections discuss end-user devices under three broad categories— mobile radios attached to vehicles, portable radios (aka hand-held devices), and consoles controlling mobile and portable radios.

11.2 MOBILE RADIOS

These devices are typically two-way radios installed on to the dash of a vehicle such as motor vehicles, motorcycles, and boats (Figure 11.1). The devices are connected to an external antenna mounted on the vehicle.

It is likely that these devices are used by officers to send voice and data messages while they are driving the vehicle. Therefore, the devices should be designed with easy to use, easy to reach, comfortable operating space, with minimum distraction.

Power is provided by the vehicle. Therefore, these devices are designed with better reception, range/coverage, and voice clarity.

Older versions of mobile radios are analog or digital simplex or half duplex radios with Push-to-Talk (PTT) buttons connected to other radios through dispatchers. More modern mobile radios are capable of full duplex communications supporting hundreds of channels with plenty of features such as external speakers close to the driver to alleviate road noise while driving. Some models also include user safety features such as "man down," which is capable of detecting whether the user is not moving and incapacitated.

They are also based on the narrowband Project 25, or Terrestrial Trunked Radio (TETRA) communications technologies, or Long Term Evolution (LTE) based broadband technology.

Figure 11.2 shows a mobile radio based on Project 25 technology. This specific one includes Wireless Fidelity (Wi-Fi) connectivity, Universal Serial Bus (USB) interface, and a commercial cellular radio interface (3G/4G) to provide the agent with various communications capabilities as well as to manage data storage [2].

Figure 11.1. A mobile radio from Motorola [1].

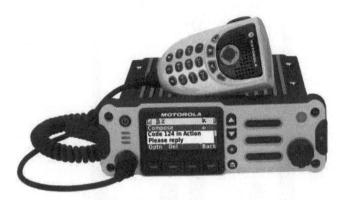

Figure 11.2. A mobile radio based on Project 25 technology [2].

Figure 11.3 shows a mobile radio based on TETRA communications technology. This specific mobile radio has a gateway and repeater functions, Bluetooth, Wi-Fi connectivity, an integrated Global Positioning System (GPS) module, and antijamming features to provide secure communications [3].

11.3 PORTABLE RADIOS

Portable radios are hand-held devices that are carried by the officers and agents on the field and other public safety personnel. These devices are equipped with a microphone and a speaker, and are powered typically by a rechargeable battery. See Figure 11.3, which shows three models—with no keypad, with a limited keypad, and with a full keypad. Most portable radios based on analog or narrowband technologies use dipole antennas. However, new portable radios based on broadband LTE technology use embedded antennas within the device.

Compared to mobile radios, portable radios are lighter, more compact, and use lower power, and thus have a smaller range and more significant mobility. They may also be equipped with Wi-Fi, Bluetooth, and GPS connectivity.

As in the case of mobile radios, portable radios also vary depending on the communications network to which they are connected. Portable radios based on the old analog technology are still in use in many parts of the world, including, surprisingly, in some regions in the USA [5].

Some vendors provide portable radios compatible with analog technology as well as narrowband technologies such as Project 25. They may support multiple frequency bands with backward and forward (Frequency Division Multiple Access [FDMA] and Time Division Multiple Access [TDMA]) compatibilities.

Project 25 technology based portable radios are currently offered by many vendors in the USA and internationally, and comply with the Project 25 protocols,

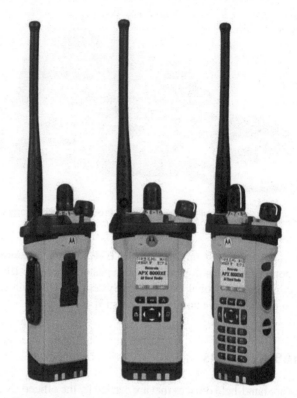

Figure 11.3. Examples of Project 25 portable radios [4].

including Over-The-Air Rekeying (OTAR), Over-The-Air Programming (OTAP), and Over-The-Intranet Programming (OTIP) [6]. See Figure 11.3 for a picture of a Project 25 portable radio. Since all vendors must comply with the Project 25 Compliance Assessment Program (Project 25 CAP) [7], which ensures interoperability, the agencies can acquire portables from any vendor.

TETRA technology based portables have similar features as Project 25 portables. See Figure 11.4 for an example of a TETRA portable terminal. Of course, the protocols, frequency bands, and compliance conditions are based on the TETRA standards specifications.

Many vendors, anticipating migration toward LTE-based public safety networks, have been busy making their products LTE compatible [9]. Also, vendors are introducing portable radios based on LTE technology (perhaps with backward compatibility to some narrowband technologies). Figure 11.5 shows a portable radio that is compatible with LTE technology.

Since LTE is a technology that is widely used for commercial services through highly popular so-called smartphones, the LTE portable radios used by public safety

Figure 11.4. An example of a TETRA portable terminal [8].

agencies will have similar features and capabilities as consumer smartphones—larger, high-resolution screens, sleek design, Wi-Fi, Bluetooth, GPS, and a wide variety of support systems and application software.

Since LTE is based on a broadband communications technology, LTE portable radios combine data, voice, and video with much higher data rates and much smaller latency. These features enable the agents on the field responding to incidents to receive and send the right information (data, voice, and video) speedily.

Also, depending on the needs of the users, some phones need to be ruggedized, glove usable, shock-, vibration-, dust-, and waterproof, and resistant to heat, drops, oil, chemicals, pressure, radiation, etc. [11]. Perhaps most importantly, these devices must support group communications, PTT and emergency buttons, man-down detection, and alarm features.

Figure 11.5. An example of an LTE portable [10].

11.4 DISPATCH CONSOLES

As discussed in Chapter 1, dispatch consoles are not end-user devices. However, they are primarily used to control and monitor many end-user devices at a single physical location. Therefore, we discuss them here. Dispatch consoles are typically located in Public Safety Answering Points (PSAPs) and dispatch centers in mission-critical communications centers. Dispatch consoles may also be located on the field in case of significant emergency situations.

Dispatch consoles help the dispatchers to communicate with the agents on the field—they help send messages, pages, connect the field agents to other networks, collect field information, etc. The agents on the field stay informed so that they can respond in real time with the right information. Therefore, dispatch consoles must set up calls and deliver emergency messages as quickly as possible. Also, dispatch consoles prioritize emergency calls to make sure that these calls go through regardless of the congestion in the network. Moreover, they ensure good voice quality so that these calls can be intelligible even during adverse conditions, in which field agents are operating.

As in the case of mobile and portable radios, dispatch consoles are also communications technology dependent. A dispatch console typically supports a microphone, a headset, and a couple of speakers. Also, the typical features of a dispatch include a built-in network monitor providing real-time feedback on the status of the network [12].

The dispatch consoles based on LTE technology present new challenges as well as opportunities for public safety agencies. Reference [13], a report prepared by the National Public Safety Telecommunications Council of the USA, lists the following requirements for an LTE-based dispatch console:

- "Interface with existing networks
- Support legacy and future technologies
- Provide robust functionality for voice, video, and data services
- To prioritize users and applications in support of an incident or event
- Provide functionality to manage network priority within an incident
- Allow seamless transfer of synchronized data sets to other locations
- Support the use of analytics to receive, organize, prioritize, and selectively route large amounts of data."

Currently, LTE-based dispatch consoles are not widely available. Several vendors are offering integrated products based on legacy systems with LTE connectivity. However, with the adoption of LTE technology by public safety agencies and other mission-critical sectors, LTE-based consoles could be readily available.

11.5 OTHERS

In addition to the end-user devices that are a part of critical communications technologies (such as Project 25 and TETRA), public safety agencies use end-user devices that are a part of commercial networks—so-called smartphones, tablets, phablets, laptops, and desktop Personal Computers (PCs) that are typically connected to commercial wireless and wireline telecom networks and the Internet through wireless (Wi-Fi) and wireline (Ethernet) local area networks.

In many cases, some of these devices (especially phones and tablets) are owned personally by the agents, and in some cases, these devices may be provided by the agencies. Personally owned devices may provide additional support and connectivity, especially for volunteer public safety agents, for example, to check their emails and apps that may be relevant to operations. However, there is also a higher security risk in allowing personal devices to be used during operations. Maintenance, loss of data, and sharing sensitive information are some of the concerns associated with these devices.

Recognizing this growing trend of so-called "Bring Your Own Device" (BYOD), many agencies around the world have been developing policies in dealing with this situation. More and more agents, especially young recruits, are already familiar with the commercial products, which are more user-friendly, faster, have a larger screen with better resolution, and more importantly, are connected to the Internet to access the worldwide web of information. The BYOD policies have been trying to take advantage of this trend by looking for ways to integrate these devices into the overall operations of the agencies [14].

As mentioned above, some of these devices may be purchased by the agency and assigned to the agents. More and more agencies are allowing agents to use these commercial devices with additional job-related applications software installed [15].

11.6 SUMMARY AND CONCLUSIONS

This chapter provided a high-level overview of the devices used by the personnel in performing public safety activities. The chapter discussed these devices under three categories—mobile radios attached to vehicles, portable radios (aka hand-held devices), and consoles controlling mobile and portable radios.

The end-user devices depend on the communications technology used in the network to which they are connected. The communications hardware and protocols are different. Also, the capabilities of the device must match the capabilities of the technology used in the network (the devices attached to narrowband Land Mobile Radio [LMR] networks [Project 25, TETRA, and Digital Mobile Radio—DMR] have limited voice, data, and video capabilities). There is no question that end-user devices compatible with LTE technology will have superior capabilities regarding performance, especially data and video capabilities.

Public safety agencies also use end-user devices that are a part of commercial networks (public wireless networks, the Internet). For example, smartphones, tablets, phablets, laptops, and desktop PCs are being used at an increasing rate to perform noncritical office applications as well as during some emergencies to supplement the public safety specific devices.

REFERENCES

1. Motorola Solutions, "CM Series Mobile Two-Way Radio." [Online]. Available: https://www.motorolasolutions.com/en_us/products/two-way-radios/mototrbo/mobile-radios/cm300d.html#tabproductinfo. [Accessed: Feb. 1, 2018].
2. Motorola Solutions, "APX™ 8500 All-band P25 Mobile Radio." [Online]. Available: https://www.motorolasolutions.com/en_us/products/two-way-radios/project-25-radios/mobile-radios/apx8500.html#tabproductinfo. [Accessed: Feb. 1, 2018].
3. Sepura, "SRG3900 Mobile Radio." [Online]. Available: http://www.sepura.com/products/tetra/mobile-radios/srg3900/. [Accessed: Feb. 17, 2018].
4. "Motorola Solutions Web Site." [Online]. Available: http://www.motorolasolutions.com/en_us/products/two-way-radios.html. [Accessed: Jul. 25, 2016].

5. Discount Two-Way Radio, "RPX4500A Public Safety Handheld Radio." [Online]. Available: https://www.discounttwo-wayradio.com/rpx4500a. [Accessed: Feb. 4, 2018].

6. Hendon Publishing, "Trends in portable radios." [Online]. Available: http://www.hendonpub.com/resources/article_archive/results/details?id=5353. [Accessed: Feb. 5, 2018].

7. Department of Homeland Security, "P25 CAP," May 22, 2016. [Online]. Available: https://www.dhs.gov/science-and-technology/p25-cap. [Accessed: Feb. 5, 2018].

8. Motorola Solutions—Europe, Middle East, and Africa, "MTP6650 TETRA Portable Two-way Radio." [Online]. Available: https://www.motorolasolutions.com/en_xu/products/tetra/terminals/portable-terminals/mtp6650.html#tabproductinfo. [Accessed: Feb. 8, 2018].

9. D. Jackson, "Motorola Solutions announces first public-safety portable radio with LTE connectivity," Urgent Communications, Mar. 18, 2014. [Online]. Available: http://urgentcomm.com/motorola-solutions/motorola-solutions-announces-first-public-safety-portable-radio-lte-connectivity. [Accessed: Feb. 8, 2018].

10. Motorola Solutions, "LEX L10 Mission Critical LTE Handheld." [Online]. Available: https://www.motorolasolutions.com/en_us/products/lte-user-devices/lexl10.html#tabproductinfo. [Accessed: Feb. 1, 2018].

11. M. Aalto, "Public safety smartphones," Dec. 29, 2015. [Online]. Available: http://lteps.blogspot.com/2015/12/public-safety-smartphones.html. [Accessed: Feb. 12, 2018].

12. "MAX Dispatch." [Online]. Available: https://www.zetron.com/en-us/products/radiodispatch/maxdispatch.aspx. [Accessed: Feb. 14, 2018].

13. National Public Safety Telecommunications Council, "Public safety broadband console requirements," Sep. 2014. [Online]. Available: http://npstc.org/download.jsp?tableId=37&column=217&id=3205&file=Console_LTE_Report_FINAL_20140930.pdf.

14. First Responder Network Authority, "Bring Your Own Device (BYOD) to the NPSBN." [Online]. Available: https://www.firstnet.gov/newsroom/blog/bring-your-own-device-byod-npsbn. [Accessed: Feb. 18, 2018].

15. FedTech Staff, "Federal agencies turn to BYOD, mobile devices in the field to attract new workers," Feb. 23, 2016. [Online]. Available: https://fedtechmagazine.com/article/2016/02/federal-agencies-turn-byod-mobile-devices-field-attract-new-workers. [Accessed: Feb. 18, 2018].

12

PLANNING FOR DEPLOYMENT AND OPERATIONS OF CRITICAL COMMUNICATIONS SYSTEMS

This chapter addresses the activities related to the development of various plans for designing, implementing, and operating critical communications systems.

12.1 INTRODUCTION

Deploying and operating a critical communications system require careful and well thought-out plans and analyses. All the immediate, intermediate, and long-term activities should be planned based on thorough investigations and studies.

Designing and implementing a critical communication system depends on many factors, including the following:

- Type of the organization—is it for a public safety agency (such as a police department) or a for-profit company (such as an oil refinery)?
- Number of organizations—is it for a single agency or multiple agencies? Is it for a single company or for multiple companies who will share the same network?

Fundamentals of Public Safety Networks and Critical Communications Systems: Technologies, Deployment, and Management,
First Edition. Mehmet Ulema.
© 2019 by The Institute of Electrical and Electronics Engineers, Inc. Published 2019 by John Wiley & Sons, Inc.

- Coverage—will the critical communication system be deployed as a local, a regional, or a nationwide system?
- Interoperability—will the critical communication system be interfaced with other agencies, other companies, and organizations, regardless of what technologies the other entities have?
- Existing deployment—is there any existing system being used? What is the underlying technology of the existing base? Is it compatible with the new system under consideration?
- Data requirements—will there be a significant need for multimedia, data-intensive applications?
- Finance—what is the level of available funds, if any? How will it be financed? Are there any issues with sharing the network with other agencies and organizations?
- Nationwide plan—is there a nationwide broadband plan? Is there any communications policy and master plan for the organization? How does a critical communications system fit into the overall plan?
- Frequency spectrum—are there frequency spectrum bands available or allocated for the new critical communications system?

Although the level of effort, activities, and contents heavily depend on the factors mentioned above, there are some common areas in planning for a critical communications system, as shown in Figure 12.1, from the project management point of view. These common areas include the management of time, cost, quality, human resources, communications, risk, procurement, stakeholders, and integration.

In a more general sense, conducting a set of feasibility studies, developing a business case, performing a risk analysis, drawing up a roadmap, developing a set of project plans, and establishing a project team as well establishing some related policies should be all part of the planning process.

12.2 DEVELOPING POLICIES

As discussed in previous chapters, leading trends in technology and plans for many countries around the world suggest that Long Term Evolution (LTE) technology should be part of any plan in deploying and upgrading critical communications systems. This does not suggest that the use of narrowband technologies should be ignored entirely. The requirements and business case may dictate the use of a narrowband technology to solve the immediate and near-term problems.

Recognizing that in the long term, critical communications systems will be based on broadband technologies, the development of an overall broadband policy for the region/state/country/organization is essential.

This is especially important and critical for the establishment of a nationwide public safety network. Equally important is the establishment of a national agency (if it does not exist) with broad powers to oversee the implementation and operation of a nationwide public safety communications system.

12.2.1 National Broadband Policy

If it does not exist already, a national broadband policy for the entire country should be developed to address commercial, public, government, and military needs and interests. The public safety network should be an integral part of this national broadband plan, which also includes the following:

- National action plans for 4G and 5G networks
- National plan to make broadband Internet accessible to all, especially in rural areas
- Related frequency allocation to support the national broadband plan
- Provisions for programs by Research and Development (R&D) and standardization for commercial companies, the government, and private institutions, and academic and research institutions and universities

See [1] for an example of the national broadband plan developed by the Federal Communications Commission (FCC) in the USA.

12.2.2 Governance Policy

If it does not exist already, a new national authority should be established to develop and manage a national program, to provide nationwide coverage for public safety communications. This entity should be given full authority to build and maintain the public safety system and make the necessary adjustments and improvement as conditions change and technology evolves.

This authority should be responsible for executing the national public safety program by leading all the efforts in detailed planning, designing, deployment, and operation of a new, national broadband public safety system. This authority should take all the actions necessary to ensure that the building and operation of the network takes place in full consultation with all the public safety agencies and first responder agencies, including central and local entities and other stakeholders such as the entity involved in spectrum allocation.

This authority will be in full charge of the new public safety system that may include some interim solutions as well as long-term broadband technology based upgrades. The following is a list of some primary responsibilities:

- Establishing all the necessary policies, including the procedures for developing Requests for Proposals (RFPs), evaluation of the submissions and the acquisition processes to build the network, and the development of operational plans, methods, and procedures such as maintenance, security, and provisioning

- Overseeing the construction, maintenance, administration, operation, and improvement of public safety networks to ensure that new and evolving technologies are taken into account
- Orchestrating with all public safety organizations and institutions to develop and use a new national public safety communication system
- Defining and executing an outreach policy to get all of the stakeholders' and vendors' community on board (including promise/process documents, info sheets, presentations, webinars, events, and local champions)
- Incorporating training, education, and evaluation programs to ensure that everyone dependent on the new system has appropriate training, facilities, equipment, and documentation as well as sufficient operational qualifications
- Interacting with national and international standards and industry organizations that have an interest in the design and operation of public safety communication systems and technologies
- Coordinating research, development, and testing of new public safety communications enhancements
- Setting up a standardization group for public safety communications to drive contributions and participation in international standards

12.2.3 Spectrum Management Policy

If it does not exist already, a nationwide policy for radio frequency spectrum management should be developed to make sure that nationwide coverage for public safety communications is an integral part of this spectrum policy. The regulatory body overseeing the frequency allocation and assignments should be engaged as early as possible. Assignment of public safety spectrum and other data communication spectrum assignments and the inclusion of an opportunity for sharing where feasible should be the primary topics to tackle.

The planned new system may require a new frequency spectrum allocation, although it is possible that the existing bands allocated for public safety may be sufficient. In any case, the regulatory body needs to be engaged to make sure that adequate frequency bands are available and allocated to the new public safety network.

The use of the 700 MHz spectrum for public safety applications is attractive because of its propagation and penetration characteristics. Additionally, a minimum of 10 MHz band for each direction is needed, especially for LTE-based networks, for transmission of video and data at broadband rates [2]. Other frequency bands or multiple frequency bands can be used as well, but it should be noted that devices need to accommodate these multiple frequencies for interoperability, which may result in higher cost.

For reliability and accessibility reasons, public safety networks traditionally are based on a reserved, licensed, and private allocation of spectrum for their exclusive use. However, due to advances in dynamic spectrum allocation techniques and cognitive software defined radio technologies, it is possible that new public safety networks may use multiple frequency bands.

With the advanced technology, it is possible that multiple users may use a shared "pool" of the frequency spectrum. LTE includes a feature called "carrier aggregation," which is an example of spectrum pooling. The full potential of spectrum pooling is available in LTE-Advanced (LTE-A). Multiple classes of spectrum holders such as commercial operators, public safety agencies, and other government/military licensees using multiple spectrum bands can cooperate in the use of a spectrum pool.

Also, the solutions to some of the requirements such as device-to-device communications [3, 4] and on-scene operations during emergencies may be accomplished through the use of an unlicensed spectrum in the 2.4 GHz and 5 GHz ranges.

A brief review how some countries deal with this issue is given below.

- Most voice Land Mobile Radio (LMR) systems constructed by public safety entities in the USA use narrowband frequencies (12.5 kHz) in the very narrow Very High Frequency (VHF) and Ultra High Frequency (UHF) bands. However, the Federal Communications Commission (FCC) has recently allocated 758–768 MHz and 788–798 MHz for base stations and mobile units use, respectively, 10 MHz wide in each direction for public safety applications. Also, for voice communications only, the 769–775 MHz and 799–805 MHz bands in 12.5 kHz narrowband increments are allocated for public safety use. The USA has also allocated a large band of the spectrum (50 MHz) in 4.940–4.990 GHz, although it is not clear how this would be used.

  In summary, in the USA, a total of 34 MHz of spectrum capacity will, therefore, be available for public safety networks within the 700 MHz band—22 MHz designated for broadband and 12 MHz allocated for narrowband communications, primarily voice [2, 5, 6].

- For civil systems in Europe, the frequency bands 410–430 MHz, 870–876 MHz/915–921 MHz, 450–470 MHz, and 385–390 MHz/395–399.9 MHz have been allocated for Terrestrial Trunked Radio (TETRA). Then, for emergency services in Europe, the frequency bands 380–383 MHz and 390–393 MHz have been allocated. These bands can be expanded to cover all or part of the spectrum from 383–395 MHz and 393–395 MHz, should this be needed.

  There is an ongoing effort in Europe to determine the most appropriate (and harmonized with other countries) frequency spectrum for broadband applications to be used by the public safety sector. Several alternatives were identified:

  o 2 × 10 MHz around 700 MHz—there are several commercial operators providing LTE services in this range. The USA approved the use of this range for public safety. There is also a tendency in other countries toward this range.

These frequencies are currently used by TV broadcasters, and it may be difficult to move these users out.

o 2 × 10 MHz below 400 MHz—this has the advantage of having longer wavelengths, and therefore provides a wider coverage with fewer cell sites. However, it is closer to the frequencies used by the military, which will be challenging to convince North Atlantic Treaty Organization (NATO) and other military organizations to move.

o Current Private Mobile Radio (PMR) frequencies in the 400 MHz area—this may allow broadband networks to be deployed quickly (without a slow regulatory process). This can only be used as a temporary solution because of its limited data capabilities and other enhanced services expected from a broadband network.

• In several countries in Asia, like in Europe, the 380–400 MHz band is reserved for public safety organizations and the military as well. The 410–430 MHz band has been allocated for civilian (private/commercial) use in other parts of the world too. In Mainland China, the 350–370 MHz band is reserved for national security networks while the first 800 band is used in Hong Kong for private networks. In Russia, 450–470 MHz is allocated for this purpose.

In Australia and parts of the Middle East, spectrum has also been allocated for public safety broadband services. For example, in Turkey, 415.5000–417.5000 MHz/425.5000–427.5000 MHz was allocated to digital featured trunk and community repeater systems and frequencies were assigned to 28 operators that were authorized to TETRA and Digital Mobile Radio (DMR) systems in 2012 [7].

An evaluation of potential spectrum alternatives to support a new public safety communications system must be performed. This evaluation should focus on the current public safety bands, including the VHF and UHF bands. Sufficient bands (at least 2 × 10 MHz) in 700 MHz or 800 MHz must be considered for public safety broadband spectrum as well. The unique attributes of each of these bands, including some technical and regulatory issues, need to be carefully considered in this evaluation. Furthermore, if and when a spectrum band can be located to support the new system, the cost of moving the existing public safety operations to this new spectrum as well as moving the existing users of this spectrum should not be underestimated and need to be calculated as part of the evaluation.

12.3 DEVELOPING A BUSINESS CASE

A business case is developed to demonstrate the benefits, alternatives, cost, and risks of the various alternative approaches identified. This is done early on so that a go/no-go decision can be made. The activities involved in developing a business case include the identification of the primary alternative solutions, various feasibility studies such as economic and technical feasibilities, and assessment of the risks associated with each alternative. A systematic comparison, a thorough evaluation of the alternative approaches, and finally making a recommendation are also part of the business case activities [8].

12.3.1 Identifying Alternatives

There are mainly three categories of alternative approaches that need to be explored—staying with the existing system, technology-specific approaches, and partnership-specific approaches.

12.3.1.1 Alternative 1—Do Nothing. Staying with the current system should always be considered as an alternative for the apparent reason—no apparent cost. While there is almost zero cost in this case, however, when the cost of the consequences of not providing advanced audio, video, and high-speed data features to public safety could be calculated, the result would be mind-boggling.

12.3.1.2 Alternative 2—Use of a Specific Narrowband Technology. This approach calls for a critical communications system based on only one of the narrowband technologies such as TETRA, Project 25, or DMR, which have been used in many countries around the world. They are mature and tested technologies. Therefore, the cost of devices and equipment is relatively inexpensive, and there are plenty of manufacturers and application developers, so it is relatively inexpensive to deploy and operate such networks (although there are slight differences regarding the cost and technology, they are not significant enough to make a significant dent in the total picture).

This approach especially makes sense if there are some agencies currently equipped with one of these technologies. Therefore, it seems logical that the existing narrowband networks can be combined and extended to make it a shared, nationwide public safety network to provide services to all public safety agencies. There is a precedence of this approach in some countries [9].

However, a narrowband-based nationwide public safety network fails to satisfy broadband capacity and quality requirements for providing advanced data and video capabilities for public safety officials.

12.3.1.3 Alternative 3—Use of a Combination of Narrowband Technologies. This approach suggests the use of two or more different narrowband technologies. This approach makes sense if there is an existing network based on a narrowband technology used by one or more agencies, and for some reason, some other agencies prefer another narrowband technology. The result will be two separate networks, which creates significant interoperability problems and a substantial increase in operational cost, as well as doubling of frequency spectrum usage.

12.3.1.4 Alternative 4—Use of a Broadband Technology. This

approach calls for a critical communications system based on LTE technology, the same technology used in commercial 4G wireless mobile cellular networks. This network will be for use only by public safety agencies. This network provides all the advanced broadband capabilities. It is expected to be relatively cheaper since LTE-based equipment, devices, and applications are in use worldwide by all wireless mobile networks, large and small. The LTE-based public safety network meets the requirement for roaming and interoperability with commercial LTE networks.

There are several persuasive arguments for why the public safety network should be a dedicated network:

- **Large-scale incidents**—the network must be available for use by first responders and emergency officials. Commercial networks can quickly become overloaded and unavailable.

- **Coverage**—nationwide coverage is essential for public safety agencies. Disasters can occur anywhere. Commercial networks cover an area only if it is profitable.

- **Backup power**—the network must be equipped with backup batteries or generators to guarantee communication during power failures. Commercial networks are limited in this category.

- **Resilience and availability**—this is a must for public safety users. Commercial networks are somewhat less concerned on this matter.

- **Security**—most public safety agencies require secure and protected systems against intruders and sabotage. Commercial networks are somewhat less concerned on this matter.

- **Less critical applications**—commercial networks can be used for less critical applications such as office systems and administrative matters.

The disadvantage of this scenario is that LTE-based public safety networks are relatively new and not mature. Although many of the features and requirements are in the standards specifications already or will be in them shortly, they are not included in the implementation of the equipment used in commercial LTE networks. So there is some uncertainty regarding when these features will be incorporated into LTE equipment. Another complication with this scenario is that it requires the regulatory bodies to allocate perhaps some new frequency bands around 20 MHz in the 700 MHz range.

The USA and some other countries are making significant progress to establish LTE-based public safety networks. This trend is essential for achieving standardized, interoperable, and "roamable" LTE-based public safety networks around the world.

12.3.1.5 Alternative 5—Use of a Combination of Narrowband and Broadband Technologies. Given the risks of uncertainty discussed earlier for a

public safety network based only on LTE, this alternative proposes a two-phased approach. In the first phase, a narrowband-based nationwide public safety network is established by building on the already-deployed network and making it available to all public safety agencies. While the effort and activities go on in expanding the existing network, in parallel, the planning and related research and development activities for building an eventual LTE-based public safety network should proceed.

The main advantage of this two-phased approach is that the network established in the first phase provides a unified public safety network initially. This first network includes many of the desperately needed features and meets many of the criteria and requirements so that all public safety agencies can take advantage of it in a relatively short period of time. Plus, since the work for building an LTE network will start immediately as well, the benefits, especially broadband video and data, that the LTE-based network will bring, will be available promptly as well. This way, two networks will be used in a complementary way—voice and low-speed data related public safety features on the narrowband network and high-speed data and video on the LTE network.

In expanding the current network, comprehensive interoperability and eventual migration of the users on the existing network to the new LTE-based network should be an essential part of the project. This approach is similar in many ways to the approach taken in the USA.

A serious concern with this approach is the additional cost of building and operating two networks.

12.3.1.6 Alternative 6—Use of Broadband Technology in Partnership with Commercial Operators.
The nationwide public safety only network approach described in Alternative 4 provides some significant financing opportunities; for example, the excess capacity can be sold to commercial operators, the utility (gas, electricity, water) companies as well as the transportation interests (buses, trucks, trains, metro, airports, marinas, ports). In other words, LTE-based public safety networks can be shared with commercial wireless cellular service providers (and possibly with some utility companies).

While this approach looks attractive from the economic point of view, it raises some concerns from the privacy, security, and performance point of view. Since commercial users may be using the same network, more stringent security and authentication capabilities must be incorporated. Since public safety networks are engineered based on "worst-case" situations, sharing it with commercial networks will increase the worst-case levels for capacity and other performance metrics, thus adding more to the cost of capital and operations.

However, because of the significant economic benefits, this approach should be seriously considered and the consequences discussed above should be carefully studied to reach optimal solutions.

12.3.1.7 Alternative 7—Use of a Commercial Broadband Network Shared by Public Safety Personnel.
In this approach, commercial LTE-based wireless operators offer broadband quality video and data services to public safety personnel. In other words, public safety agencies arrange with one or more commercial LTE operators for the provision of mobile broadband services.

Another variation in this solution is that public safety agencies may operate as Mobile Virtual Network Operators (MVNOs), where the public safety agency makes a bilateral business agreement with a commercial operator to have dedicated access for public safety personnel.

In this scenario, there is no need for a separate network just for public safety purposes. This option has distinct cost advantages. The only expense is the service fee to be paid to the commercial operator.

However, there are serious concerns with this approach:

- Commercial networks are not designed and engineered to guarantee many critical requirements (e.g. security, reliability, availability, and long power-backups), which are necessary for public safety applications. Network access and reliability are especially required during times of emergency. Therefore, at critical times, commercial public networks cannot offer the resilience, speed, and unique features that are critically important for first responders.

- Geographical coverage of commercial networks is limited to the areas where they make economic sense. Public safety requires nationwide coverage.

12.3.2 Feasibility Studies

A feasibility study is an analysis of the viability of an idea or a project through a disciplined and documented process of thinking [10]. It is perhaps the most critical phase in the development of a project [11]. Feasibility studies are conducted before developing a formal business plan [12]. A feasibility study results in a feasibility report that provides documentation that the idea was thoroughly investigated. The feasibility report explains in detail whether the project under investigation should be carried out. A typical feasibility study includes technical feasibility, financial/economic feasibility, operational feasibility, market feasibility, organizational/managerial feasibility, environmental feasibility, and legal feasibility. Note that not all these types may apply to a given feasibility study [12].

Before a feasibility study begins, we need to know what is being studied. Is it for upgrading an existing system or establishing a brand new system? What are the alternative approaches? The previous section addressed this issue in detail. Once the alternative approaches are identified, various feasibility studies need to be conducted on each of these alternatives.

It is also essential that a set of criteria addressing necessity, attainability, completeness, consistency, and complexity must be established as the first step in a feasibility study. These criteria define the conditions for the selected approach to be acceptable to the users and other stakeholders.

A set of more specific criteria that are directly related to the technology, operations, cost, finance, and the users must be established as well. Some examples are user expectations, data rate requirements, performance (throughput, capacity, and latency), availability, reliability, resiliency, security, scalability, evolvability, interoperability, manageability, cost-effectiveness, and more.

In addition to the feasibility studies, a risk analysis on the alternative approaches needs to be performed. As a first step, potential risks need to be identified by answering a simple question—what can go wrong if we choose this alternative? The next step is to assess the possible impact of each risk. Finally, the risk analysis includes a discussion and identification of the steps that may have to be taken to avoid or minimize each risk [8].

12.3.3 Interoperability Concerns

Another critical factor in evaluating the alternative approaches is interoperability. The recommended solution may include the deployment of two separate networks. Furthermore, there may be existing network(s) that may need to coexist with the new solution either temporarily or for longer terms. Therefore, interoperability among the newly deployed networks as well as with the existing systems is among the paramount interests.

Furthermore, interoperability with commercial systems is essential as well, especially during emergencies. Since the emergency call number system is one of the primary triggers for public safety activities, the recommended scenario must be interoperable with emergency call centers as well.

12.3.3.1 LTE Interoperability. As discussed earlier, an LTE-based network may not have some features, like direct mode operations, available immediately. Therefore, additional measures must be taken to provide interoperability between the LTE-based system and the existing PMR/LMR systems. There are several vendors such as Etherstack that offer solutions to provide complete interoperability [13]. The solution requires the deployment of some gateways in the core network to handle a set of supplementary service features based on public safety PMR network call types. These include group calls, fast call setup times, emergency calls and encrypted voice services, Push-To-Talk (PTT)-over-LTE features as well as providing interfaces to Project 25 or DMR based systems. Also, several vendors offer solutions for the interface to the Internet, and to wireline public telephone networks.

In the case of Project 25, there is support for Telecommunications Industry Association (TIA) TR.8 Inter Sub Systems Interface (ISSI) and Console Subsystem

Interfaces (CSSI), which are essential for interoperability, from several vendors already. A major push in the Project 25 community is for the development of specifications (such as P25 PPToLTE) for interoperability between Project 25 and LTE public safety networks. Since the ISSI is based on Internet Protocol (IP)/Transmission Control Protocol (TCP) standards including Session Initiation Protocol (SIP) and Real-time Transport Protocol (RTP), which are also included in the LTE standards, the interoperability between these two should be relatively straightforward.

Some of the vendors also include, in their gateways, the voice coders/decoders (vocoder) used in DMR and Project 25 networks, which transmit voice calls without transcoding. This approach results in higher quality and no extra delays, most importantly allowing end-to-end encryption between a PMR end device and an LTE end device.

12.3.3.2 Backward Compatibility.
If a new system cannot interoperate with previously deployed equipment, the active operations of the new system will be in jeopardy. This creates severe barriers to the acceptance and use of the new system. This is not an easy problem to solve, especially when wireless systems are involved. This may mean the new system must accommodate different radio implementations, which increase the cost. In general, this might be ameliorated if the new system can achieve automatic detection and adaptation. There may be other solutions as well; for example, a fall back to simple features, disabling the advanced ones for a while.

One key advantage of Project 25 over TETRA is that it is backward compatible with the standard analog Frequency Modulation (FM) radios (this also allows a phased introduction of the Project 25-based network). This backward compatibility of Project 25 also helps interoperability when personnel from another, different system roam into a Project 25 system [14].

DMR is also backward compatible with the existing analog systems while there's no backward compatibility in TETRA.

12.3.3.3 Interoperability with Commercially Deployed Systems.
There are substantial benefits for the public safety networks interoperating with commercial networks. Some of these benefits are listed below:

- Commercial networks can be used during emergencies.
- The public safety network can provide leasing services to commercial operators for additional revenues.
- The public safety network built in areas that are not of interest to an existing commercial operator can be shared via bilateral agreements:
  - One possible arrangement is called "Radio Access Network (RAN) sharing," where all of the participants (public safety network and one or more

participating operators) provide the same geographic coverage through the hosting RAN (in our case, the public safety network owns the hosting RAN).

o Another possibility is to provide a roaming arrangement. When roaming, a commercial subscriber uses the visited network (the public safety network) when outside of the home network geographic coverage and within the visited network's geographic coverage.

- The commercial equipment might be used to serve during an emergency.
- New apps can be developed on smartphones that can interwork with public safety equipment, especially where the user is in relatively benign environments.
- Experimentation and prototype deployment for testing and evaluation for the public safety network can be done on a commercial network before it is incorporated into the public safety network.
- A commercial network can be used for training public safety officials.

The Project 25-based standards have provisions for interoperating with commercially deployed systems. Next Generation Technology Independent Interoperability of Emergency Services (GERYON) is a European Union (EU) funded project to develop solutions across diverse PMR networks (e.g. DMR) and public mobile networks (e.g. LTE/WiMAX/Wi-Fi) based upon the IP Multimedia Subsystem (IMS) [15].

12.3.3.4 Interoperability with Emergency Call Centers.

A national public safety network design needs to take into account emergency call centers for obvious reasons (the number 911 in the USA and the number 112 in Europe are designated for emergency calls). The 112 system is a primary trigger mechanism for public safety operations. These call centers need to be integrated into the public safety network so that emergency number call takers can contact appropriate dispatchers in public safety agencies to trigger appropriate actions. Traditionally, 112 calls are made from fixed wired phones, but increasingly, these calls come from mobile phones that can be located through their proximity to specific base stations and, in many cases, the use of Global Positioning System (GPS).

Public safety networks should also take into consideration (and take advantage of) technologies such as the Internet of Things (IoT) and Machine-to-Machine (M2M) communications. The Internet-enabled sensors may generate or signal these calls with an excellent location accuracy to trigger public safety operations automatically.

Another critical area in interoperability is that all the agencies using different narrowband networks based on the same technology must be able to interoperate among themselves [16]. This is true for the LTE-based network as well, but since the LTE network will be developed from the ground, this aspect should be part of the network design.

12.3.4 Comparison of Alternatives and the Recommendation

Once the alternatives are identified and analyzed, the last step is to recommend one of the options. All those feasibility studies carried out for each of the alternative approaches provide a solid base for comparing and eventually identifying the best alternative to recommend for implementation and deployment. There are several models used in the comparison. Table 12.1 provides a high-level comparison of the alternative approaches previously discussed against primary concerns. Note that attainability refers to whether the underlying technologies are available, attainable, and implementable in a timely manner, technically and otherwise. Completeness refers to whether the alternative provides a comprehensive and complete solution and meets all user demands, and accommodates all the current and anticipated users.

Once a high-level comparison similar to Table 12.1 is made for each alternative approach, the next step is to select three or four top candidates and perform a more comprehensive comparison on them by using more quantitative financial and scoring models, perhaps combining quantitative and qualitative criteria [9]. Typically, the

TABLE 12.1. A Qualitative Comparison of the Alternative Solutions

Alternatives	Criteria				
	Broadband	Interoperability	Attainability	Completeness	Cost
1 Do nothing	No	No	N/A	No	N/A
2 Use of a specific narrowband technology	No	Yes	Yes	No	Low
3 Use of a combination of narrowband technologies	No	Yes	Yes	No	Medium
4 Use of a broadband technology	Yes	Yes	Yes	No	Medium
5 Use of a combination of narrowband and broadband technologies	Yes	Yes	Yes	Yes	High
6 Use of broadband technology in partnership with commercial operators	Yes	Yes	Yes	No	Low
7 Use of a commercial broadband network shared by public safety personnel	Yes	Yes	Yes	No	Low

TABLE 12.2. A Sample Scoring Model [9]

Criterion		Weight	Alternatives		
			A	B	C
Financial	• ROI	15%	2	4	10
	• Payback	10%	3	5	10
	• NPV	15%	2	4	10
Strategic	• Alignment with strategic objectives	10%	3	5	8
	• Increased market share	10%	2	5	8
Organizational	• Likelihood of achieving project's objectives	10	2	6	9
	• Availability of skilled team members	5	5	5	4
Project	• Cost	5	4	6	7
	• Time to develop	5	5	7	6
	• Risk	5	3	5	5
Customer	Customer Satisfaction	10	2	4	9
	TOTAL	100	2.65	4.85	8.50

Note: Risk scores have a reverse scale—that is, higher scores for risks imply lower levels of risk.

scoring models require assigning weights and scores to each criterion. The weights and scores assigned are subjective, and experience may help create a more realistic business case. Table 12.2, based on a different project (not related to critical communications systems), is provided here to show a sample scoring model for comparing three alternative solutions. As shown, each criterion is assigned a relative weight, and a score (between 0 and 10) for each alternative. The total score of an alternative is simply a weighted average of all the individual scores for each criterion. The table shows that alternative C scores the highest and should be selected as the option for this project.

Based on comparisons similar to the above methodology, one of the alternatives is selected as the most appropriate approach to implement.

In the case of critical communications systems, among the alternative approaches identified previously (Table 12.1), it is clear that what is needed in the long term is an LTE-based, converged solution for a mobile broadband critical communications system. This alternative provides strong resiliency and better Quality of Service (QoS), virtualization, synchronization, and enhanced security. Also, LTE provides broadband data and high-quality graphics, image, and video, as especially needed for a nationwide public safety network. Only an LTE-based network can provide public safety agencies with advanced multimedia capabilities with the speed and low latency demanded.

Some standards and development work related to public safety features are required for LTE networks to reach the ideal and comprehensive public safety functionality and service levels. Therefore, some immediate and intermediate steps may be taken to realize the long-term objective of a common, converged LTE-based dedicated public safety network. A narrowband technology based network may be considered to serve as a "safety net," providing mission-critical voice features while the LTE-based network catches up with voice-specific functionality. This intermediate step implies that the apparent cost will probably be higher. However, narrowband technologies offer options to re-use the existing site and facilities, which include the re-use of site shelters or buildings, antenna masts, antennas, cabling, and the transmission network. This reduces the capital expenditure. Narrowband technologies also provide other options to migrate toward an LTE-based public safety network, which again is expected to lower the cost significantly.

12.4 DEVELOPING PROJECT PLANS

Once one of the alternative solutions is selected, the follow-on steps mainly include designing, implementation, and testing and deployment to make it ready for the agencies to use the system. However, before all these steps are taken, there needs to be a set of plans, called project plans, to define, monitor, and control a detailed set of activities in each phase of the project. All these are part of the project management phases and activities.

The application of knowledge, skills, tools, and techniques to project activities to meet project requirements is the primary management goal of any project, including high technology projects such as putting together a critical communications system. Figure 12.1 shows the typical project management areas, each of which requires a separate project plan to identify and monitor individual activities in that management area.

Each of these plans should provide unambiguous, measurable, trackable answers to the following questions:

- What work needs to be done?
- Who will do the work?
- What resources will be needed to do the work?
- When will they do the work?
- How long will it take?
- How much will it cost?

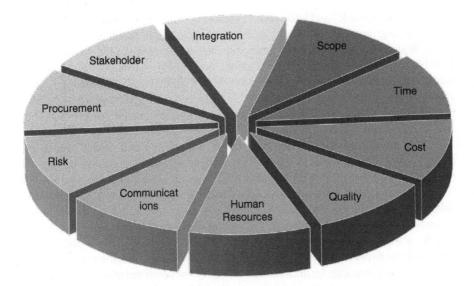

Figure 12.1. Project management knowledge areas [17].

The project must have a governance structure within the boundaries of the governance policy discussed in Section 12.2.2. This project-specific governance structure must define the following:

- Structure of the project organization
- Authorization related to this project
- Oversight and accountability
- Decision-making process for this project.
- Resources (staff, materials, tools, etc.) needed for this project

A project charter containing all these project plans, including the project's governance structure, the project's measurable objectives, and explicit commitment to the project, must be produced. The project charter serves as an agreement and as a communication tool for all of the project's stakeholders.

Moreover, perhaps, most importantly, a project team led by a qualified and able project manager with extensive experience and a proven track record in managing the deployment of critical communications systems should be hired. The rest of the project team should be established consisting of people with technology skills, business/organization knowledge, and interpersonal skills, among other qualifications.

In determining the plans, especially the ones that involve scheduling and resources, it is essential to keep in mind that building a public safety network that is nationwide, flexible, and adaptable to changing needs is a complex undertaking. Therefore, a gradual, iterative, and incremental approach to building the network is highly desirable. This approach requires that each increment be tested under the most possible realistic conditions by the stakeholders involved, before the next increment is handled. Next, in the case of nationwide network deployment, the establishment of a framework for a national test bed is highly recommended. This testbed can be used to validate the implementation of a new system, subsystem, or component before they are put into service.

Once the critical communications system is deployed and turned over to the users, it needs to be monitored continuously to make sure that it performs the way it is designed. Related to this objective, there are numerous activities, collectively called "network and service management." This is more commonly known as Operations, Administration, and Maintenance (OA&M) of the network and services. Although it is not part of the design, implementation, and deployment of the project, OA&M-specific activities need to be planned as well, ideally in parallel. This is traditionally done after the network is in place, which results in undesirable situations including all kind of features that may be difficult to manage. However, if the plan is developed in parallel to the other planning activities, the OA&M plan will provide valuable input to the design and implementation of the network as well.

12.5 SUMMARY AND CONCLUSIONS

This chapter discussed various plans for designing, implementing, and operating critical communications systems.

Although the level of effort, activities, and contents heavily depends on many factors, there are some common areas in planning for a critical communications system. More specifically, conducting a set of feasibility studies, developing a business case, performing a risk analysis, drawing up a roadmap, developing a set of project plans, and establishing a project team as well establishing some related policies should be all part of the planning process.

Recognizing that future critical communications systems will be based on broadband technologies, the planners should develop an overall broadband policy as well as a policy for radio frequency spectrum management. Also, a dedicated national authority with broad powers to oversee the implementation and operation of a nationwide public safety communications system is essential.

A business case should be developed to demonstrate the benefits, alternatives, cost, and risks of the various alternative approaches identified. The business case should include several feasibility studies (economic and technical) and a comprehensive evaluation of the alternative approaches, and a recommendation.

Finally, project plans must be developed to define, monitor, and control a detailed set of activities in each phase of the project.

Deploying and operating a critical communications system require careful and well thought-out plans and analyses; therefore, all the developments and deployment should be based on thorough investigations and studies.

REFERENCES

1. Federal Communications Commission, "Connecting America: The national broadband plan," Mar. 2010. [Online]. Available: https://transition.fcc.gov/national-broadband-plan/national-broadband-plan.pdf.

2. "Spectrum management for the 21st century: The President's spectrum policy initiative," National Telecommunications Information Agency, Mar. 2008.

3. P. Janis, C.-H. Yu, K. Doppler, C. Ribeiro, C. Wijting, K. Hugl, O. Tirkkonen, V. Koivunen, "Device-to-device communication underlying cellular communications systems," *Int. J. Commun. Netw. Syst. Sci.*, vol. 2, no. 3, pp. 169–178, 2009.

4. J. Liu, Y. Kawamoto, H. Nishiyama, N. Kato, N. Kadowaki, "Device-to-device communications achieve efficient load balancing in LTE-Advanced networks," *IEEE Wireless Commun.*, vol. 21, no. 2, pp. 57–65, Apr. 2014.

5. "Public Safety Spectrum Trust." [Online]. Available: http://www.psst.org/. [Accessed: Jul. 18, 2016].

6. Federal Communications Commission, "FCC takes action to advance nation-wide broadband communications for America's first responders," News Release, Jan. 2011.

7. The Information and Communication Technology Authority, Republic of Turkey, "ICTA 2012 Annual Report," 2012.

8. J. T. Marchewka, *Information Technology Project Management: Providing Measurable Organizational Value*. John Wiley & Sons, 2003.

9. B. Deverall, "Briefing for E-Gif Working Group on incorporating the APCO P25 standards into the e-government interoperability framework," Added Value Applications, Sep. 19, 2006.

10. H. Hoagland and L. Williamson, *Feasibility Studies*. University of Kentucky, 2000.

11. H. Tompson, "Business feasibility studies: Dimensions of business viability," Best Entrepreneur, 2003.

12. W. Truitt, *A Comprehensive Framework and Process*. London: Quorum Books, 2003.

13. Etherstack, "PMR-LTE network solutions: Push-to-talk PMR over LTE," White Paper, 2013.

14. J. Oblak, "Project 25 Phase II," EFJohnson Technologies, Dec. 2011. [Online]. Available: https://www.efjohnson.com/resources/dyn/files/688228za6510319/_fn/PROJECT+25+PHASE+II.pdf.

15. European Commission, "Next generation technology independent interoperability of emergency services," GERYON Final Report. [Online]. Available: https://cordis.europa.eu/result/rcn/164396_en.html. [Accessed: Sep. 14, 2018].

16. P. R. Kempkerhttp, Department of Homeland Security, "Basic gateway overview." [Online]. Available: http://www.c-at.com/Customer_files/Fairfax%20rally/Module%201%20-%20Basic%20Gateway%20Overview.ppt. [Accessed: Sep. 13, 2018].

17. Project Management Institute, *A Guide to the Project Management Body of Knowledge (PMBOK® Guide)*, 6th ed. Project Management Institute, 2018.

13

ECONOMIC AND FINANCIAL CONSIDERATIONS FOR DEPLOYING CRITICAL COMMUNICATIONS SYSTEMS

This chapter provides an overview of economic and financial considerations for deploying a critical communications system, including general cost and financing related components and the alternative financing forms. Although most of this discussion applies to all kinds of critical communications systems, our focus in this chapter is on the public safety networks used by public safety agencies, which could be local, regional, or nationwide. Therefore, deploying a public safety network is primarily a *public project* sponsored by a local, regional, or a national government.

13.1 INTRODUCTION

Deploying a public safety network requires substantial financial investment. Accurate cost estimations and appropriate financing are critically important. Estimated costs of implementing and operating nation-wide public safety networks are running over many billions of dollars. For instance, the total estimated cost of the proposed public safety network in the USA is expected to be in tens of billions of dollars [1].

Fundamentals of Public Safety Networks and Critical Communications Systems: Technologies, Deployment, and Management,
First Edition. Mehmet Ulema.
© 2019 by The Institute of Electrical and Electronics Engineers, Inc. Published 2019 by John Wiley & Sons, Inc.

In this chapter, we focus on the public safety networks used by public safety agencies, although many financing options discussed here also apply to critical communications systems used by corporations. Government projects significantly differ from corporate or private projects. Public projects are not usually profit driven and are undertaken for the good of society, with an objective to raise the welfare level of the public. Therefore, most public projects are financed by the taxpayers and are not subject to the classic cost–benefit analysis performed by corporations. However, in general, it is a good policy to implement corporate financing rules as much as possible to ensure that the most suitable project is accepted and the costs are minimized while the benefits are maximized, so as to not waste the taxpayers' money in inefficiencies.

Financing a nationwide public safety network must address the initial investment mostly involving Capital Expenditures (CAPEX) and perhaps, more importantly, the ongoing Operational Expenditures (OPEX). Therefore, it is necessary to consider and integrate all possible forms and sources of financing.

One approach is to share the public safety network with non-public safety users who pay fees for sharing it. This approach is essential since sharing, in general, is a long-term agreement, and the cash flows obtained from a partnership do not only contribute to the initial capital expenditures but also support the ongoing operations.

The US and European Union (EU) proposals also show that the initially required funds should come from a government budget. However, no government can easily include a line item in billions of dollars without proper financial planning and preparation and without disturbing the ongoing operations. The government must consider the use of all possible forms of financing including government bonds, equipment leasing, vendor financing, private partnerships, asking in-kind contributions from utility companies, and the like.

13.2 COST AND BENEFIT OF DEPLOYING AND OPERATING A PUBLIC SAFETY NETWORK

A good understanding of the cost structure of deploying a public safety network would be valuable to understand the factors affecting the cost. Similarly, the types and amount of tangible and intangible benefits need to be understood not only to determine the amount of money required for the network to be financed, but also to evaluate various alternatives.

13.2.1 Cost of Deploying and Operating a Public Safety Network

Cost estimations of deploying and running public safety networks vary significantly [2], and the accuracy of those estimations are hard to assess due to unclear methods and assumptions.

In cost estimation, it is essential to consider the Total Cost of Ownership (TCO) for designing, developing, acquiring, deploying, maintaining, and supporting the network over its useful time. The TCO calculation should include the following:

- *Direct cost or upfront cost*—the initial cost of designing, building, installing, and testing
- *Ongoing costs*—salaries, training, and upgrades
- *Indirect costs*—loss of productivity, time lost, and others

In estimating the direct cost, Reference [3] suggests that the number of cell sites required should be the primary focus since the deployment and operating costs of a network are proportional to the number of cell sites. US studies to determine the cost of a nationwide public safety network have estimated that to cover 98% of the population, about 40,000 towers are needed. These US studies also highlight that there are about 3 million first responders in the USA, but in a shared public safety network with the inclusion of commercial partners, the system will have approximately 30 million users [3]. It is, therefore, reasonable to assume that economies of scale work and produce much efficiency, such as the use of lower cost devices.

Determining the cost in the case where the government partners with an outside commercial firm is entirely different. In this case, public safety agencies share the network with commercial subscribers.

- *Coverage*—For commercial systems, the areas where paying customers reside and travel are the priority. However, for public safety systems, all the areas must be covered since emergencies may occur anywhere. Since the number of cells required is highly dependent on the total area covered, one has to determine first the area currently covered by the existing system (if there is one) and then compute the additional cell sites required to cover the rest of the areas.
- *Capacity*—Commercial systems are typically designed to optimize capacity, whereas public safety networks are over-engineered so as to give peak performance. Furthermore, public safety responders may require higher data rate applications like high-resolution live video. In other words, the key motivation for a public–private partnership is based on the fact that public safety communications systems are designed for worst-case capacity scenarios. Therefore, the number of cell sites required for each proposal must be carefully studied for a variety of capacity requirement scenarios.
- *Other requirements*—Moreover, commercial cellular users and public safety responders have different requirements such as the availability of a wireless signal within buildings or the reliability of a signal in the coverage area, so in a shared public–private network, it may be necessary to determine where on that continuum a network planner should design. Understanding those issues

ensures a better network design and not a compromise that is unnecessarily expensive for commercial users and inadequate for public safety [4].

Typically, the number of cells required in a wireless network is dependent on the size of the cells in the network, and the size of a cell is dependent on the frequency and bandwidth of the spectrum used. As the frequency increases, the size of a cell tends to decrease; however, as the bandwidth is increased, the capacity that a cell can support increases.

Also, there are some other factors that need to be included in the calculations of the cost. Some of these are direct whereas some are indirect costs. For example, the cost of capital needs to be accounted in these figures.

13.2.2 Benefits of Deploying and Operating a Public Safety Network

As in the case of costs, it is also important to look at the big picture in estimating the benefits. In the project management context, this is defined as the Total Benefits of Ownership (TBO), which should include direct, ongoing, and indirect benefits over its useful time on the network [5].

As discussed previously, the objective of building and operating a public safety network is to raise the welfare level of the public. It is not usually profit driven and is undertaken for the good of society. Therefore, quantifying the benefits of implementing and operating a public safety network is not straightforward.

In addition to not so tangible, indirect, socioeconomic benefits, there may be actual, direct benefits as well, such as improving the accuracy and efficiency by reducing errors and duplications, improving decision making by providing timely and accurate information, improving user service by making the system faster and more convenient to use, and perhaps obtaining revenues via sharing the network with one or more partners.

Since public safety networks are always built with excess capacity and operated using only a fraction of the available capacity, partnership is not only possible but also recommended by several studies to use this reasonable cash-generating opportunity [6, 7]. A visible and logical partnership would be with utility companies. In support of this argument, one may use the US data reported in 2012 showing that utility companies in the US spent over $3.2 billion on their telecommunications [6, 7].

Another critical issue is the opportunity cost of using the spectrum for public safety networks instead of selling/auctioning it to commercial service providers and obtaining economic gain for the government [8]. For example, the annual consolidated socioeconomic benefits computed for a sample set of EU countries with a total combined population of 300 million people are estimated to be around 21 billion Euros [9] (this study takes into account the reduction in labor force, crime reduction, mobility reduction, increase in real estate values, etc. due to the availability of a public safety network). The opportunity cost of the above scenario for the EU sample set with a population of 300 million is to sell the spectrum at auction. The government can obtain a one-off economic gain totaling 3.7 billion Euros, based on an MHz cost per pop benchmarked from international sales of 700 MHz and 800 MHz and the United Kingdom (UK) results (an MHz cost per pop "refers to one megahertz of bandwidth passing one person in the coverage area in a spectrum license" [10]). This example shows that the benefits are several times greater than the opportunity cost, suggesting that deploying the public safety network is more beneficial to the society.

13.3 FINANCING ALTERNATIVES

In this context, financing is an activity that is aimed at finding the money to build and operate a public safety network. The larger the required initial investment expenditures, the more complex is the financing activity. There are some approaches to financing a government project. The following is a brief discussion on three popular financing alternatives.

13.3.1 Bond Financing

Bond financing, also called "borrow-and-purchase," is a classic financing technique, where the government borrows the required amount from creditors/lenders.

More substantial amounts, in general, require standardized contracts that can be offered to many lenders, typically fixed-income investors, without negotiating with them separately. The terms of the arrangement such as the agreed return and duration, payment frequency with the lenders, and the risk level (perhaps assigned by a rating agency) must be part of the standardized contract.

The government may finance the project by using one of (or a combination of) the bond financing methods listed below.

- **Fixed interest rate contracts**—should be preferred if the inflation and interest rates are not expected to go down in the future
- **Adjustable interest rate contracts**—are preferred if the inflation and interest rates are expected to go down in the future
- **Cancelable contracts**—are preferred if the cancellation can be made at the government's discretion and, if possible, without penalty
- **Perpetuity contracts**—are bonds with no expiration date, and the principal will never be paid back to the creditor, but the creditor receives a fixed interest payment forever [11]
- **Convertible debt contracts**—enable the borrower to start with an interest rate that is lower than a comparable standard bond and offers the lender the option to convert the bond into an attractive profit sharing arrangement, if and when the operations generate profits

13.3.2 Lease Financing

Lease financing is similar to vendor financing—instead of dealing with creditors, the government asks the lessor (e.g. the equipment vendor) to cover the cost of the project. Leasing is considered a popular alternative to the borrow-and-purchase option, and is especially preferred for larger projects.

Lease financing is mainly preferred by non-public safety entities (tax paying) since the entire lease payment is tax-deductible (in many countries) for the lessee while only the interest is tax-deductible in case of the "borrow-and-purchase" financing.

Since the government is not a taxpayer, there will be no tax-related advantages from the leasing. However, the government may offer tax breaks to certain parties in return for their involvement in refinancing the public safety networks.

There are some different leasing arrangements. We discuss the following three major lease types:

- **Operating leases**—these include both financing and maintenance. Ordinarily, operating leases require the vendor to maintain and service the leased equipment. These types of leasing typically have a built-in cancellation clause. In general, operating leasing provides valuable operational flexibility, which should be preferred even if the cost of leasing is the same or even a bit higher than the bond financing.
- **Capital leases** (aka financial leases)—these do not include maintenance service and are not cancelable. However, the vendor receives the leasing payments equal to the full price of the equipment plus a return on the invested capital. Typically, the government negotiates the price with the vendors, and then arranges to have a leasing company to buy the equipment from the vendor, and simultaneously execute a leasing contract.
- **Hybrid leases**—these present the features of both financial leases and operating leases. Deploying a public safety network is a good candidate for hybrid leases due to its size and complexity.

13.3.3 Financing via Sharing the Network

This alternative financing approach requires a partnership with one or more private (non-public safety) entities to share the burden of financing as well as the use of the deployed network. Because large-scale emergencies do not take place all the time, it is possible for the commercial partners to share some of the public safety spectra to serve commercial subscribers. Also, in case of rare emergencies, the partnership is designed in such a way that the public safety partner is allowed to access both the public safety and the commercial spectrum.

This approach creates side income/cash flows by charging public partners an agreed fee for utilizing the network. Similarly, a public–private partnership can be established so that a commercial partner commits to serving public safety agencies in return for access to valuable spectrum in which it could serve paying customers [11].

When this financing method is considered, the government would want assurance that the private party's use of the network will not by any means lower the quality and availability of the network for its intended use and compromise the safety and confidentiality of the public safety network.

It is expected that public safety networks generate a significant amount of income by sharing their capacity with other parties so that they may efficiently finance themselves.

13.4 EVALUATION OF FINANCING ALTERNATIVES

There are several mathematical methods to evaluate the economics of a given project. The following provides a brief discussion of some of these methods:

- **Net Present Value (NPV)**—determines the present value of the initial investment and subsequent cash flows at a required discount rate [12]. The formula for calculating NPV is:

$$NPV(i, N) = \sum_{t=0}^{N} \frac{R_t}{(1 + i)^t}$$

  where
  t is the time of the cash flow
  i is the discount rate, which depends on the risk structure of the firm (the treasury bond rate for government projects)
  R is the net cash flow
  N is the total number of periods

  If NPV is positive, the project is considered to be viable.
- **Internal Rate of Return (IRR)**—determines the rate at which the project breaks even. The IRR is the discount rate that makes the NPV equal to zero [13]. The implication of this is that if the IRR is less than the deployment cost of capital (in percentage), the project under consideration is not viable.
- **Modified Internal Rate of Return (MIRR)**—is a variation of the IRR, and includes the best parts of NPV as well. MIRR determines a rate of return with an explicit reinvestment rate assumption. In other words, MIRR equates the cost of the initial investment to a terminal value that is computed by using the cost of capital as the reinvestment rate of the cash flows [13].

Table 13.1 provides a comparison of these four methods, which need to be carefully evaluated and integrated as much as possible to maximize the benefits and

TABLE 13.1. Popular Investment Viability Methods [13]

Method	Issues if the Method is Selected
Net Present Value (NPV)	NPV does not produce a percent return but its reinvestment rate is correct, and there are no multiple solution possibilities. It is the most common method, is well understood by the industry, and is academically trusted.
Internal Rate of Return (IRR)	IRR assumes the reinvestment rate is the project specific IRR, creating potential estimation biases. Under certain circumstances, IRR may produce multiple solutions, which confuses the evaluators.
Modified IRR (MIRR)	Very reliable and may be used in any feasibility analysis without issues. It is much more complicated than the other procedures and is now well understood in the industry.

minimize the cost of the proposed project. The government needs to use its negotiation power as a tax collector when dealing with private partners. In other words, any borrower considers future tax reliefs as the return on their investments since this process is similar to various bond financing applications.

Most projects produce simple future cash flows as the single return. However, projects such as public safety networks are much more complicated and produce material cash flows as well as socioeconomic benefits that are very difficult to combine, integrate, and quantify.

Simple steps of feasibility analysis may be listed as follows:

- Collect information and estimate the initial investment outlay.
- Estimate the future cash flows generated by the initial investment during the life of the project.
- Compute the cost of capital components. If multiple financing forms are used, calculate the weights and component costs accordingly, including tax-related adjustments.
- Compute the Weighted Average Cost Of Capital (WACC). This is merely the percentage cost of financing the capital investment.
- Compute the project's NPV using the WACC as the discount factor to see whether the final result is positive so they can proceed with the project as it is feasible on the corporate scale.

13.5 SUMMARY AND CONCLUSIONS

Deploying and operating a public safety network requires substantial investment. Financing such an investment is a complicated issue. The first step should be to estimate the expected initial amount needed to deploy a public safety network.

An accurate cost computation for deploying a public safety network is near impossible. Specific mathematical models can be useful and should be used for cost estimations. This estimate provides a reference value, a benchmark to compare various financing alternatives, which need to be carefully evaluated and integrated as much as possible to maximize the benefits and minimize the cost of the proposed project.

Once the required amount is estimated, the government should seek the most effective way to finance it.

- If the government runs a budget surplus and funds are available without canceling another essential project (i.e. no opportunity cost), then there is no need for any additional action to finance the initial amount.
- If the government budget has no funds earmarked for the project, then generally imposing a new tax or initiating an arrangement with a third party are the most common ways to raise the initially required funds.
  - o One standard method used to promote specific projects is to implement an ad valorem tax on cigarettes—an insignificant earmarked tax will not only help accumulate the required initial funds but will also institutionalize the steady cash flow that will contribute to the operating budget of the project in the future years.
  - o When dealing with private partners, the government can and should use its negotiation power as a tax collector; any borrower would consider future tax reliefs as the return on their investments. This approach is similar to bond financing applications, where the bondholder, the creditor, or the vendor considers future taxes in place of interest returns. Data from the USA suggests the involvement of utility companies as a partner in financing and sharing public safety networks. This partnership creates efficiency for both minimizing the cost and maximizing the benefits.

REFERENCES

1. Federal Communications Commission, "FCC takes action to advance nation-wide broadband communications for America's first responders," News Release, Jan. 2011.
2. Federal Communications Commission, "The public safety nationwide interoperable broadband network: A new model for capacity, performance and cost," White Paper, Jun. 2010.
3. R. Hallahan and J. M. Peha, "Quantifying the costs of a nationwide broadband public safety wireless network," in *Proc. 36th Telecommun. Policy Res. Conf.*, Arlington, 2008.
4. R. Hallahan and J. M. Peha, "The business case of a nationwide wireless network that serves both public safety and commercial subscribers," in *Proc. 37th Telecommun. Policy Res. Conf.*, Arlington, 2009.

5. J. T. Marchewka, *Information Technology Project Management: Providing Measurable Organizational Value*. John Wiley & Sons, 2003.
6. J. M. Peha, "A public–private approach to public safety communications," *Issues Sci. Technol.*, vol. 29, no. 4, 2013.
7. Utilities Telecom Council, "Sharing 700 MHz public safety broadband spectrum with utilities: A proposal," A Proposal, Oct. 2012.
8. TETRA and Critical Communications Association, "Spectrum saves lives: High speed data for the police and emergency services," presented at the The Operators Workshop, 2011.
9. A. Grous, "Socioeconomic value of mission critical mobile applications for public safety in the EU: 2 × 10 MHz in 700 MHz in 10 European countries," Centre for Economic Performance, London School of Economics and Political Science, Dec. 2013.
10. S. Hansell, "Verizon licks its cheap megahertz pops," *Bits Blog*, Mar. 21, 2008. [Online]. Available: https://bits.blogs.nytimes.com/2008/03/21/verizon-licks-its-cheap-megahertz-pops/.
11. "Consol (bond)," *Wikipedia*, Jan. 1, 2018. [Online]. Available: https://en.wikipedia.org/wiki/Consol_(bond).
12. A. Gallo, "A refresher on net present value," *Harvard Business Review*, Nov. 19, 2014. [Online]. Available: https://hbr.org/2014/11/a-refresher-on-net-present-value. [Accessed: Mar. 1, 2018].
13. N. Bolari, "Indirect returns and use of NPV in financial viability modelling of critical communications networks," *IEEE Commun. Mag.*, vol. 54, no. 3, pp. 38–43, Mar. 2016.

14

DESIGNING, IMPLEMENTATION, AND INTEGRATION

This chapter addresses the design, implementation, testing, and integration of a critical communication system. When these activities are completed successfully, the network is ready for service.

14.1 INTRODUCTION

Once the business case including the recommended approach is accepted, the next logical steps are to develop a detailed network architecture and design, procure and develop the components, and install them. Finally, it is necessary to perform various types of field work, which typically involves performing unit, system, and integration tests.

14.2 NETWORK ARCHITECTURE AND DESIGN

This phase starts with developing a high-level *network architecture*, followed by detailed low-level *network design*. The typical goals of network architecture and

Fundamentals of Public Safety Networks and Critical Communications Systems: Technologies, Deployment, and Management, First Edition. Mehmet Ulema.
© 2019 by The Institute of Electrical and Electronics Engineers, Inc. Published 2019 by John Wiley & Sons, Inc.

design are to determine the coverage, capacity, Quality of Services (QoS), performance, and security to meet the requirements. It is essential for the high-level architecture to have a high level of acceptance from all stakeholders.

Usually, a network architecture includes three primary areas—a *functional architecture* where the functions necessary and needed for the network and a *physical architecture* where the functional entities are mapped into corresponding physical counterparts; the operational principles and procedures, as well as the data formats used in its operation, may be part of these architecture specifications as well.

Once the high-level architectural components and concepts are documented, the next step is to develop and prepare a low-level network design. This detailed design document is a blueprint describing every single detail that is necessary to build, implement, and install the network. The detailed design should include, for example, the specification of the ports, cables, connectivity (which port on a given component is to be connected to which other component, with what kind of cable/wire), power, and location.

There are some similar design considerations between a cellular network and a large trunked critical communication network. For example, both types of networks have base stations and radio towers, radio access, backhaul, and core network equipment. So the planning and deployment should have similar issues to address.

At the same time, due to performance-related requirements, commercial networks and public safety networks differ significantly in addressing geographical coverage and traffic capacity in their design. For example, critical communication networks are more concerned about coverage. Usage traffic patterns are also entirely different.

Luckily, general network architectures for popular critical communications technologies are well specified and documented. A rather brief discussion on them is provided below in Section 14.2.1. Similarly, the architecture and design aspects of an LTE-based broadband network are also well known and mostly standardized. Section 14.2.2 addresses the architectural and design aspects of LTE networks.

14.2.1 Designing Narrowband Technologies Based Network

The network architecture for narrowband-based networks is similar especially to the ones based on TETRA and Project 25.

Figures 14.1 and 14.2 show a general deployment view of a Project 25 based critical communications network and a TETRA-based critical communications network, respectively. These figures identify the major components and interfaces of the respective technologies. Chapter 5 provides a more detailed description of the Project 25 based network architecture. Similarly, Chapter 6 provides a more detailed description of the TETRA-based network architecture. In the following paragraphs, a brief synopsis of these architectures is provided for the sake of completeness of this chapter.

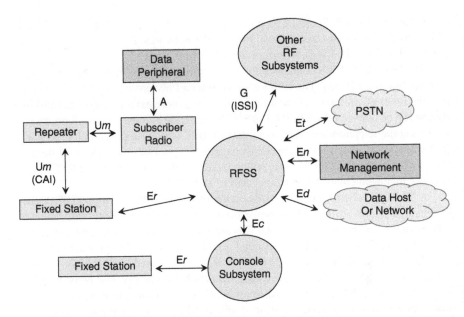

CAI, Common Air Interface; ISSI, Inter Sub System Interface; RF, Radio Frequency; RFSS, RF Sub System; PSTN, Public Switched Telephone Network

Figure 14.1. Project 25 general system model [1].

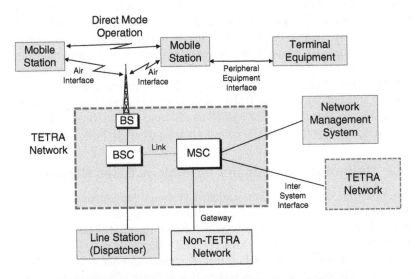

Figure 14.2. Deployment view of TETRA standard elements.

Figure 14.1 shows Project 25's physical high-level system architecture, which is similar to the one we discussed above for TETRA networks. Project 25 interfaces are employed to form the RF Sub-System (RFSS) infrastructure.

The Project 25 RFSS contains a collection of base stations that support the Common Air Interface (CAI). The RFSS also includes all necessary control logic to support the intersystem interfaces and protocols. For a more extensive system, multiple RFSSs can be connected via ISSIs. Project 25 systems are connected to telephone networks via the telephone interconnect interface (Et), which supports both analog and ISDN telephone interfaces. Network management system(s) are supported through the network management interface, which facilitates the management of RFSSs with network management system(s) connected to an RFSS.

A TETRA network contains a collection of trunked Base Stations (BSs), each of which supports a tower of transmission antennas communicating with two-way mobile radios and portable radios. All BSs are connected to one or more Base Station Controllers (BSCs). Sometimes a BS and a BSC could be in one product called the Site Base Station (SBS). The BSCs (and SBSs) are connected to a Mobile Switching Center (MSC), also called the Switching Control Node (SCN). The Network Management System (NMS) connected to an MSC is used to monitor, administer, and maintain all the components mentioned above. SBSs can also support wired connections to Line Stations (LSs) used as dispatch consoles. A TETRA network may be connected to other TETRA networks via Inter System Interfaces (ISIs), which connect the MSC of a TETRA network to another TETRA network. A TETRA network may be connected to a non-TETRA network through the MSC via gateways as well [2].

In the case of DMR-based critical communications, the network design is comparably simple. Chapter 7 provides a more detailed description of the DMR tiers. Here, a brief refresher is provided. Tier I architecture is mainly composed of end-user devices (e.g. mobile stations), without the use of repeaters. There is no infrastructure; no telephone interconnects. Tier 2 architecture includes repeaters (base stations) without any trunks among them, and DMR nodes providing switching and control functions. Tier 3 is a trunked architecture. DMR sites can be connected via an IP-based core network. Also included in the Tier 3 DMR architecture are gateways, dispatcher consoles, and network management stations, as in the case of TETRA architecture [3].

14.2.2 Designing a Broadband Technology Based Network

As discussed on several occasions earlier, LTE-based broadband commercial networks have been around for a while, and therefore many aspects of their architectures and designs are specified in a large number of standards documents published by the 3rd Generation Partnership Project (3GPP) standards body [4]. However, the use of the same technology in building critical communications systems, especially a nationwide public safety network, is relatively new. Some architectural variations for public safety needs have been discussed already in previous chapters. Whether the

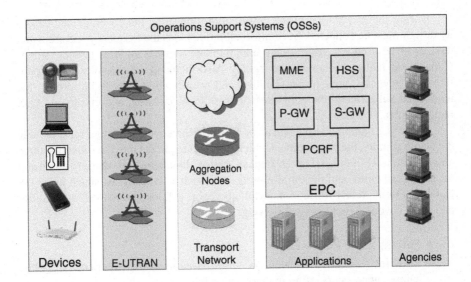

E-UTRAN, Enhanced Universal Mobile Telephone System Radio Access Network; HSS, Home Subscriber Server; MME, Mobility Management Entity; PCRF, Policy and Charging Rules Function; PDN, Packet Data Network; S-GW, Serving Gateway; P-GW PDN, Gateway

Figure 14.3. LTE physical architecture.

network is shared with other not for profit partners or with commercial partners will introduce architectural variations. At this stage, it is assumed all these alternatives have been evaluated, and one of the options is selected.

Regardless of how the network will be shared and operated, an LTE-based network has a similar functional architecture, as discussed in more detail in Chapter 8. The physical architecture and actual deployment of such a network will have a different look and will have different physical components and segments. Figure 14.3 shows a possible deployment view identifying several segments of a deployable network.

These segments are e-UTRAN, transport, Evolved Packet Core (EPC), applications, and operations support systems. In the following sections, design aspects of each of these segments are discussed in detail.

14.2.2.1 Radio Access Segment. The LTE radio access segment, called the evolved UTRAN, includes numerous Evolved Node B (eNB), LTE base stations, dispersed geographically all over the country. Therefore, determination of the number of the eNBs and their locations is of paramount importance in the design of the network. The design of this segment consists of three primary phases—the preparation phase, the coverage and capacity design phase, and the site selection phase.

During the preparation phase, information impacting the coverage and capacity design must be collected. Such information includes the frequency spectrum and devices to be used by public safety communities (devices such as laptops, smartphones, and tablets generate more traffic). Also included are the roles of the public safety agencies, their responsibilities, and prioritization, user traffic profile, the number of subscribers, the geographical areas to be covered, and end-to-end QoS requirements.

If this is a nationwide public safety network, the coverage design must include the whole country, the geography of which can be categorized into the following three areas: (i) densely populated urban areas in metropolitan cities, (ii) densely populated areas in small cities, and (iii) rural-remote areas such as villages. As stated in the US National Public Safety Telecommunications report [5], some other critical requirements for coverage include the following:

- Specific traffic load levels
- Calls for service and incident locations based upon the experience of public safety agencies operating in that geography
- Sensitive and critical infrastructure such as schools, shopping malls, event venues, transportation corridors, and other infrastructure that may be subject to significant public safety incidents or terrorist attacks

Several different methods can be used to predict radio coverage. The most accurate one is field measurements by using RF planning tools and simulators to monitor coverage and capacity; for example, the ratio of signal loss in a building or inside a vehicle in a particular area. The measurement models include the characteristics of the selected antenna, the terrain, and land use. The designers must also use a digital geographical map and a database to store all the information collected from simulations and RF measurement to plan the cell sites. The results of all RF measurements and simulations provide useful information including antenna directions, handover parameters, quality and strength of signal, and uplink and downlink throughput, which help designers to determine the cell site locations, the number of cell sites, and cell types. The cell types can range from macrocells serving thousands of users to femtocells serving small facilities or offices ($\sim$10 users). The locations of the cell sites must also be carefully examined by considering seismic events (which is a must in countries with a history of earthquakes), wildland fires, flooding, and wind and ice storms.

The LTE-based public safety network should leverage as much as possible the existing cell sites (and the infrastructure) where it makes engineering and economic sense. The decision of whether an existing site can also be used for the LTE network depends on the size of the facilities and whether the cell site (location) is the most suitable according to the network design, coverage, and cost requirements.

14.2.2.2 Core Segment. The LTE core network, called EPC, provides packet routing, mobility management, and service provisioning concerning the user profile, location, and traffic load. The EPC consists of Mobility Management Entity (MME), Home Subscriber Server (HSS), Serving Gateway (SGW), Policy and Charging Rules Function (PCRF), and Packet Data Network Gateway (PGW), as shown in Figure 14.3. The public safety LTE core also includes security infrastructure to fulfill the security requirements set by public safety agencies.

There are some key factors to be examined and considered during the design phase of the EPC, based on the overall private LTE deployment strategy. One of the critical decision points is whether to deploy a centralized or a distributed core network architecture. While the centralized approach provides a single point of operations and control, the distributed model will improve the redundancy and thus the resiliency of the network and services. For a public safety network, a distributed model is recommended only because of the importance of network availability. The other key decision point is to determine whether the SGW and PGW will be deployed into a single node or separate nodes. It is recommended that a single node including both SGW and PGW functions should be seriously considered in the case of a private network. The main advantage of this model is small latency. The MME pool, which consists of some MMEs, is also another factor impacting the design. The MME pool provides redundancy and load balancing to meet increasing signaling in the network. The LTE core network design also includes core network dimensioning, which is used to determine the number of nodes and the capacity required. The critical dimensioning parameters are the number of transactions per second to support subscriber mobility, the number of subscribers, and throughputs.

14.2.2.3 Transport Segment. A backhaul network is a transport facility, which is deployed between the E-UTRAN and the EPC to carry traffic from the eNBs to the elements in the core network, as shown in Figure 14.3. Backhaul networks must support secure communication, sufficient bandwidth, and higher traffic rates to meet the critical requirements of the LTE-based public safety network [6]. The backhaul design consists of a high-level design first, followed by a low-level design.

The primary focus of the high-level design is to determine the transport media, transport technology, and the topology. The selection criteria are based on the evolving LTE requirements, such as higher capacity, latency, and QoS. To meet the LTE requirements, the following are recommended: fiber optics as the transport media, Layer 2 Ethernet as the transport technology, and the ring as the topology. A high-level design of a mobile backhaul network based on the ring topology is shown in Figure 14.4.

The fiber optics provide higher capacity, the use of Layer 2 is more straightforward and cheaper compared to Layer 3, and the ring topology provides carrier-class transportation with capabilities such as fast link failure detection and fast recovery, which are crucial in a public safety network.

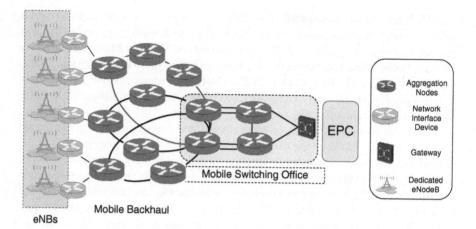

Figure 14.4. An example of backhaul high-level architecture.

The **high-level design** should also include the number of aggregation nodes and their locations, which are determined based on the capacity and distance between the hubs, handover procedures to the Mobile Switching Office (MSO), high-level service profile, bandwidth profiles of the users, and failure scenarios and detection mechanisms. The MSO is a physical facility that typically hosts some aggregation nodes and the components of the EPC.

The primary focus of the **low-level design** is to determine the critical parameters such as the port and Virtual Local Area Network (VLAN) assignments, wavelengths, the interconnection of aggregation nodes, and the network interface devices. A detailed QoS design should also be included to fulfill the Service Level Agreements (SLAs) requirements such as one-way delay, two-ways delay, frame loss, and frame loss ratio during data packet transmission.

14.2.2.4 Other Segments.
In addition to the radio access, core, and transport segments, there are application and network-management related components of an LTE network, as briefly discussed below.

Applications—Applications considered to be used by public safety agencies are very important for the design and planning of the LTE network and its sites. Because some applications require high bandwidth capacity while others involve real-time transmission, they must be taken into consideration when designing the network. The typical applications that could be used in the public safety sector are video, dispatch, fingerprint, image transfer, voice over IP, push to talk, mobile database query, and monitoring related applications.

Operation Support Systems (OSS)—OSSs are the critical components used in a public safety network to keep the network up and running, providing its services satisfactorily, as promised. The OSSs can be centralized or distributed depending on the network size and capacity. The category of OSS functionalities includes fault management, performance, configuration, security, and inventory management.

14.3 IMPLEMENTATION AND INSTALLATION

Once the planning and design phase is completed and approved, the activities in implementing or acquiring and installation of the components and systems of the network begin. Like the design phase, this phase also includes careful considerations for each segment of the overall network. Before the actual installation begins, a deployment plan must be developed. The *deployment plan* should include installation, configuration, and test procedures, and the timeline for the nodes as well as for the whole network to be deployed.

Radio Access Network (RAN) deployment consists of cell site housing, installation of radio base stations (eNBs), integration, and site acceptance testing. A detailed deployment plan is required to coordinate the installations promptly and to perform the necessary configurations simultaneously. The cell site housing and the equipment to be installed in a disaster area, such as an area where an earthquake or a wildfire has occurred, must be earthquake and fire resistant and have physical strength and thermal protection to continue the operation despite unfavorable environmental conditions. Power is another crucial factor for the availability of cell sites. The radio base stations must also be equipped with battery backup units to keep the operation going in case of a loss in commercial power.

A Method of Procedure (MOP) must be prepared and used in the installation and configuration of the radio base station. A MOP must include all the necessary instructions, such as prerequisites and precheck of the system, loading software, powering eNBs, site-specific configuration scripts, and exit criteria. Each eNB is connected to an OSS. After installation, to check the eNB's visibility in the network, an acceptance test must be performed to verify node configuration.

The mobile backhaul is composed of a suite of components based on the technology and the topology determined in the design phase. Network Interface Devices (NID) and aggregation routers are two common primary nodes of a backhaul, as shown in Figure 14.4. These aggregation routers must be equipped with hardware and software redundancy, power redundancy, scalability up to 100 Gbps, and management and open interfaces, and support Metro Ethernet Forum (MEF) compliant services and security services. The sites should be ready and network fiber built before the installation and configuration of the nodes.

Even though core network deployment is more straightforward compared to RAN deployment, there are still some key factors to be considered. All core nodes must have redundant hardware, highly scalable architecture, and a carrier-class platform supporting high transaction rates, activations, tracking paging, and handoffs. A key deployment consideration is the integration of multiple core functions on a single platform. MMEs must be configured to support load balancing and congestion. An acceptance test is performed to validate the QoS, connections between the nodes, e.g. testing the S1 interface between eNodeB and MME, quality of experience (e.g. video stream), subscriber profile, and handoff.

Several key factors impact the OSS deployment strategy. These factors include hardware and capacity dimensioning based on the number of cell sites and the number of nodes to be deployed in the network, footprint, power and space requirements, and centralized or distributed architecture. High Availability (HA) requires that the equipment and its components be continuously operational for a desirably extended length of time, cluster or replication, and backup options as manual or automated. In addition to other OSSs, a centralized OSS is recommended to view and manage the end-to-end network.

14.4 SYSTEM INTEGRATION, VERIFICATION, AND VALIDATION TESTING

The public safety network is most likely to be deployed by using equipment from multiple vendors. While the transport network equipment is acquired from one vendor, the other vendors can supply equipment for radio access and core networks. Therefore, it is crucial to perform end-to-end system integration and verification testing to verify the requirements of public safety authorities [7].

A steering committee made up of all stakeholders should be set up to provide guidance, including processes (for example how to track and resolve the issues found), roles and responsibilities, entry–exit criteria, and test objects. When the system integration is completed, a verification testing is performed to verify stability concerning Key Performance Indicator (KPI) measurements. Also tested are the media quality such as video or speech quality, robustness (e.g. link brakes), maintainability (fault, performance, and configuration management), capacity and coverage (uplink, downlink, and throughput), and end-to-end performance (e.g. delay, service access time).

System integration and verification testing may be subdivided into the following categories:

- Site Acceptance Testing (SAT)—it is conducted to ensure that the network and system elements and support facilities are installed as specified in the architecture and engineering specifications so that Network Verification Tests (NVT) can begin. Also, SAT assures site readiness from operational perspectives not customarily included or covered in the NVT tests.
- Network Verification Testing (NVT)—it is performed to validate significant design features such as service connectivity, data rate limiting, and protection switching functions. The strategy used for network verification is to test network element interaction with the objective of exercising data networking and path connectivity, reliability, redundancy, and synchronization.

- Operational Readiness Testing (ORT)—it is performed to verify the KPIs and the alarms in case of a failure, such as link down and node failure; it is performed after the nodes have been installed and configured per the design document. The ultimate objective of ORT is to make sure that the elements are correctly in place and working efficiently and can support the network's intended operations. ORT is conducted in the key areas, such as fault management, configuration management, and performance management.

14.5 SUMMARY AND CONCLUSIONS

This chapter provided a brief discussion on the design, implementation, testing, and integration of a critical communication system.

A high-level *network architecture*, followed by a detailed low-level *network design*, should be developed. A network architecture includes a functional architecture, a physical architecture, and operational principles and procedures. The low-level network design is a blueprint describing every single detail that is necessary to build, implement, and install the network.

Even though there are some similar design considerations (radio, transport, core, etc.) between a cellular network and a large trunked critical communication network, the design of public safety networks differs significantly from that of commercial networks in addressing geographical coverage and traffic capacity in their design.

Designing narrowband-based networks is relatively simple compared to designing LTE-based broadband networks. Luckily, LTE-based broadband commercial networks have been around for a while, and therefore many aspects of their architectures and designs are widely available. However, the use of LTE in deploying critical communications systems is relatively new, and some architectural variations may be necessary for public safety needs, as discussed in previous chapters.

Before the actual installation begins, a deployment plan, which includes installation, integration, and test procedures for the nodes as well as for the whole network, should be developed. The components of sizeable public safety networks are likely to be from multiple vendors. Therefore, end-to-end system integration and verification testing are necessary to make sure that the system is operational and fully satisfies the requirements of public safety authorities.

REFERENCES

1. S. Burfoot, P. Chan, and D. Reitsma, *P25 Radio Systems Training Guide, Revision 4.0.0.* Codan Radio Communications, 2013.
2. J. Dunlop, D. Girma, and J. Irvine, *Digital Mobile Communications and the TETRA System.* John Wiley & Sons, 2013.

3. "ETSI TR 102 398: DMR general system design." [Online]. Available: http://www.etsi.org/deliver/etsi_tr/102300_102399/102398/01.03.01_60/tr_102398v010301p.pdf. [Accessed: Jul. 10, 2017].
4. "3GPP." [Online]. Available: http://www.3gpp.org/. [Accessed: Jul. 24, 2016].
5. National Public Safety Telecommunications Council, "Defining public safety grade systems and facilities—final report," A NPSTC Public Safety Communications Report, May 2014.
6. Aviat Networks, "Five recommendations for building FirstNet-ready backhaul networks," White Paper, Aug. 2013.
7. D. Witkowski, "Effectively testing 700 MHz public safety LTE broadband and P25 narrowband networks." [Online]. Anritsu Company, 2013. Available: https://www.anritsu.com/en-us/test-measurement/solutions/en-us/effectively-testing-700mhz-public-safety.

15

OPERATIONS, ADMINISTRATION, AND MAINTENANCE OF CRITICAL COMMUNICATIONS SYSTEMS

This chapter provides a discussion on the operations, administration, and maintenance of critical communications systems. This area is also known as "network and services management," which is well documented, especially for commercial information and telecommunications networks and services. The purpose of this chapter is to provide a short introduction to the field and focus only on its application to critical communications systems.

15.1 INTRODUCTION

Once a critical communications system is installed and ready for agencies to use, the system needs to be monitored continuously to make sure that it provides the features and services as designed without any disruption and any degradation in its performance. In addition to preventive measures, there must be capabilities to detect and repair any fault and anomalies with minimum disruption of the communications service. Just a few seconds in critical emergency situations make a big difference in saving lives.

Fundamentals of Public Safety Networks and Critical Communications Systems: Technologies, Deployment, and Management, First Edition. Mehmet Ulema.
© 2019 by The Institute of Electrical and Electronics Engineers, Inc. Published 2019 by John Wiley & Sons, Inc.

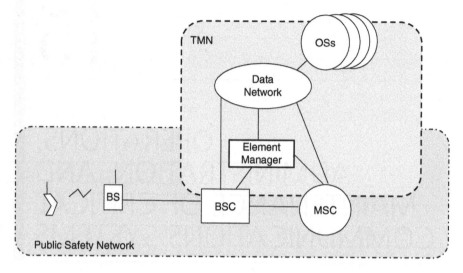

BS Base Station   OSS Operations Support System
BSC Base Station Controller   TMN Telecommunications
MSC Mobile Switching Center   Management Network

Figure 15.1. Network management infrastructure.

The network and its components are typically built with the necessary network and service management features, as discussed earlier in the chapters that are dedicated to narrowband and broadband technologies. In many cases, the related standards also include specifications related to management of the networks as well as standardized network management and control interfaces and protocols (such as the Simple Network Management Protocol [SNMP]) for external support systems.

In addition to the public safety network itself, there is a separate "network and service management infrastructure," dedicated to the maintenance, operation, administration, and provisioning of the actual network. This separate infrastructure, traditionally called the Telecommunications Management Network (TMN) [1–3], includes a set of Operations Support Systems (OSSs) connected to each other as well as to the components of the critical communications system, as illustrated in Figure 15.1.

The network management infrastructure is a framework that also includes applications, plans, policies, procedures, and people to provide a variety of functions under the five management functional areas—fault management, performance management, configuration management, security management, and accounting management [4]. Also, the TMN identifies five hierarchical logical layers when discussing these functional areas, as shown in Figure 15.2.

Briefly, the business management layer includes functions related to business aspects such as profitability, policies, goal settings, service and feature definitions,

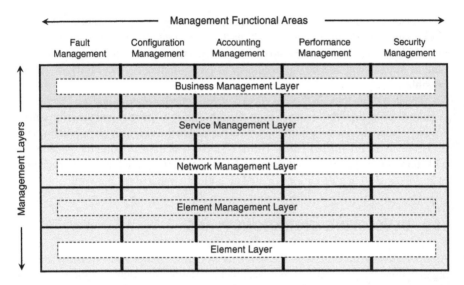

Figure 15.2. TMN layers and functional areas.

legal and regulatory issues, etc. The service management layer includes functions related to the services that the network offers. These functions include service request, service creation, service provisioning, end-to-end service management, Quality of Service (QoS), Service Layer Agreements (SLAs), and billing. The network management layer includes functions related to the resources that make up the total network. The functions at this layer are mainly responsible for receiving aggregated and summarized data from the element management layer and creating an end-to-end view within its domain. The element management layer includes functions related to a group of, perhaps, similar network elements. This layer is also called the subnetwork management layer. Finally, the element layer includes those management functions (within the element) related to a specific network element [4, 5].

In general, configuration management is responsible for designing, installing, testing, initializing, and operating network and service resources. Fault management deals with network anomalies and failures in real time. Performance management is responsible for providing reliable and high-quality network performance and maintaining the quality of service for users. Security management is responsible for protecting against threats that may compromise network resources, services, and data. Finally, accounting management is responsible for managing and administering user related matters (e.g. billing), and includes functions related to the use of resources, and the appropriate reports [4].

Before we go further, let's make it clear that the size of the infrastructure and the number of layers, OSSs, etc., depend on how extensive the critical communications system is. For a small system, one simple OSS may be sufficient, but for an extensive nationwide public safety network, many OSSs may be necessary.

Network management is critically important, especially for public safety networks, since in extreme situations, for on-the-scene operations, the network must be incredibly resilient and must be up continuously. That demands the utmost care, continuous monitoring, and taking necessary steps to intervene if necessary as soon as possible.

The first thing to do in this area is to develop a master plan, called the Operations, Administration, and Maintenance Plan (OA&M) plan, to identify and describe what is needed to keep the network humming, as explained in the next section (sometimes this is just called the *operations plan*). The next step is to acquire or build the systems that will support the operations and develop detailed policies, guidelines, and procedures to be followed to handle various related activities to manage the network.

15.2 DEVELOPING OPERATIONS PLANS

An *operations plan* describes the supports systems, capital and human resources, organizations, responsibilities, policies, and operation procedures required to monitor and manage the network efficiently. The operations plan is mostly a long-range master plan that drives further effort in developing detailed requirements for support systems and developing detailed policies and procedures to be used by the network management people. Depending on the size of the network, there could be separate plans for maintenance, traffic engineering, configuration/provisioning, and security. However, here in this book, we refer to them collectively as the operations plan.

The plan should begin with an overview of the critical communication system, its subsystems, and its components. Then the plan should identify operations and maintenance requirements for each of these physical components and each subsystem [6]. The plan should also identify logical entities such as various data that need to be collected and managed. In general, an operations plan should include in high level:

- Identification of operations support systems, operations support tools, operation support applications, and their high-level requirements
- Identification of operations centers (aka network control centers) and their high-level functions and roles
- Workflow procedures to coordinate the activities of different centers and organizations
- Responsibilities for operations and management areas
- Security policy, for example how often the passwords can be changed, access to resources and applications
- Privacy policies, restrictions on specific data
- System health-related procedures, how to monitor and report system health and maintenance actions
- Policies on system backup—daily, weekly, or monthly backup, automated or manual backup

- Data collection and archiving, what data and for how long they are to be stored; for example, how long the alarms are stored—a month or a year
- Escalation procedures in the event of emergencies
- Network monitoring procedures, parameters to be used, how to collect and report performance management data
- Fault management procedures, classifications of alarm severities, trouble ticket and escalation procedures for alarms, e.g. Tier 1 (first responders), Tier 2 (technical expertise), Tier 3 (vendor's technical expertise); eliminate single points of network failure.
- Change management policy when adding a new service or equipment
- Asset management policy, accurate and updated information on all assets related to operations
- System upgrades policy, downtime, and fallback procedures

The operations plan should be prepared with the utmost care since it serves as the blueprint used to develop the network and service management infrastructure. The plan needs to be prepared with close interactions with all the stakeholders, and their input must be incorporated into the plan. Most importantly, the final plan must be accepted by all the stakeholders, especially by the people who are involved in managing the network.

The detailed requirement specifications of the OSSs identified in the plan must be prepared by expert groups, and the activities for the implementation and acquisition of these systems as well as their installations and testing must follow.

The policies and procedures identified in the operations plan must be developed in detail, including the operations center(s) and the operations people involved. The operations plan and procedures are executed by staff members of the operations and control centers, which must be set up to monitor and control the end-to-end network or regions to manage all the segments of the network. These detailed procedures should describe, among other things, the flow of activities and the data necessary for the staff in these centers or the staff in the field to perform the activities.

When multiple centers and regions are involved, the procedures and models used by each center must be aligned with international standards such as the Telecommunications Operations Map (eTOM) [3] defined by the Telecommunications Management Forum (TMF) and the Information Technology Infrastructure Library (ITIL) [4].

15.3 OPERATION SUPPORT SYSTEMS (OSSs), TOOLS, AND APPLICATIONS

The term "*Operations Support System*" (OSS) is used traditionally by large telecommunications. The term "*Network Management System*" (NMS) is also used to refer to the systems used in managing (i.e. operating, administering, and maintaining) networks. NMSs are frequently used within the context of data networks, although it should not be surprising to find literature that uses these words (OSS and NMS) interchangeably.

15.3.1 Many Types of OSSs: Layered Organization

Depending on the size of the network, there could be many OSSs—regional, national, specialized (e.g. billing), general purpose, etc.—to provide end-to-end network service as well as process support. Many OSSs may be based on a hierarchy where low-level OSSs may be directly connected to the network elements and collect data from them, and in turn, forward the collected information to higher level OSSs for further specialized processing, as illustrated in Figure 15.3.

For relatively small networks, vendors tend to offer solutions that provide all the functions integrated into a single NMS, or distributed throughout various modules that may be connected through an IP network via the client-server architecture [7, 8].

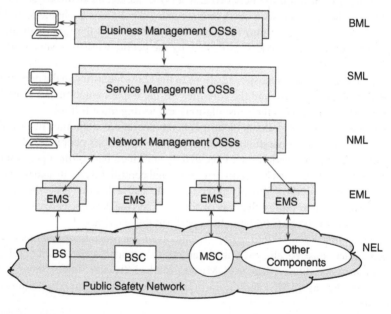

BS Base Station   MSC Mobile Switching Center
BSC Base Station Controller   NEL Network Element Layer
BML Business Management Layer   NML Network Management Layer
EML Element Management Layer   OSS Operations Support System
EMS Element Management System   SML Service Management Layer

Figure 15.3. Hierarchical architecture of OSSs in large networks.

If the goal is to deploy a nationwide public safety network, there need to be one or more common nationwide centers and associated systems used by all the public safety agencies in a coordinated way. Some forms of these may already exist, but they may need to be updated to keep up with new technology.

In TETRA, the OSSs at various hierarchical layers (e.g. local network management systems and central network management system) are connected to each other via the I5 interfaces running the SNMP [9]. A central network management system, sometimes called a Manager of Managers (MoM), provides a complete overall graphical view of the whole network by filtering, correlating, and consolidating duplicate data corresponding to events generated by multiple sources across the network. The MoM approach provides better visibility, faster determination of the causes, and therefore faster repair of the faulty components [10].

15.3.2 Interfaces among OSSs

OSSs must support standard open interfaces and multivendor and multitechnology solutions and applications, and have real-time performance data processing and intelligent troubleshooting capabilities to resolve network-related issues promptly.

In the context of critical communications systems, OSSs are used to support network and service management functions. Project 25, TETRA, and LTE technologies, which are used in critical communications systems, all have standardized interfaces to their OSSs [11].

For example, in Project 25, the interface referred to as En is dedicated to network management related interchange between an NMS and the nodes in the network (note that Project 25 and TETRA standards prefer the use of the term NMS rather than OSS).

The SNMP is the preferred protocol standard chosen by critical communications network technologies [12]. As discussed before, the SNMP is used as a protocol between the NMS and network elements (components of critical communications systems) to exchange network and service management data related to network traffic, alarms, provisioning, etc. This exchange of information is typically initiated by the NMS, such as requesting status information or usage data collected at the components. Alarms are the exceptions; a component upon detecting an alarm reports it automatically to the NMS [13].

15.3.3 OSSs Supporting Network Management Functions

OSSs record all the network events including the date, time, source, event time, and all the information relating to mobiles and groups in use on the network, including static (e.g. mobile ID, group ID) and dynamic data (e.g. location, status). The recorded and stored data (in various administrative/operational databases) are used to prepare various reports such as traffic load, usage, and performance. The backup and automated

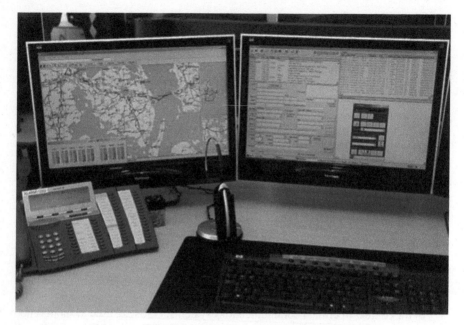

Figure 15.4. Components of an alarm monitoring system.

mass update to new software versions of all nodes in the system are also essential features of OSSs.

From the monitors attached to OSSs, users, dispatchers, and gateways can be managed and maintained by using various tools, typically via graphical user interfaces. These tools are user-friendly and flexible for configuration, maintenance, administration, and performance optimization of the network. It is vitally important that, in critical communications systems, these user interfaces be intuitive, user-friendly, and flexible for easy access to configuration and control of the entire network. Many equipment providers also include language selection features where the system can be configured for the desired language.

A summarized list of status and alarms are automatically displayed for all elements in the network for further analysis. The positions of all the components with their status are also displayed on a map [14]. These topographic views give immediate notification of network status and alarms (Figure 15.4).

15.3.4 Tools and Applications Supporting Operations

Network management functions may be supported by the OSSs via the tools and applications embedded directly in them or via external systems, which may be linked to the OSSs to download data for further processing (if the external system is not directly linked, the data may be downloaded in some other ways such as via external storage devices). These tools are intended to simplify a variety of network and service management functions such as network configuration, root cause analysis, and network optimization. Some example applications and tools used in a P25-based network are given below [15].

- Unified Network Configurator (UNC)—allows configuration of all radio network and transport devices
- User Configuration Manager (UCM)—allows configuration and control of consoles and subscribers
- Radio Control Manager (RCM)—allows monitoring and management of radio faults

In general, there are numerous applications and tools in the market for monitoring critical communications systems, collecting data, analyzing them, and generating reports. Another valuable tool, especially useful for small to medium size critical communications systems, is the Terminal Management System (TMS), which provides automated remote configurations, feature activation, software maintenance, and software upgrades of the terminals [16].

15.3.5 OSSs Supporting Network Technologies

The OSSs may be technology specific. An EMS is built to provide element management layer functions for a component of a TETRA network, or a Project 25 network, or an LTE network. Furthermore, each EMS is vendor-specific as well. Therefore, for example, a TETRA technology-based network may have components from several different vendors, each of which provides its own product-specific EMS.

Similarly, an NMS is built to provide network management layer functions for the whole network. NMSs are network specific. In other words, an NMS for a TETRA network provides functions specific to TETRA technologies. An NMS may be built by a network equipment vendor or by an independent vendor.

Some aspects of EMSs and BMSs may be more technology independent. In other words, for example, the majority of a billing system is network technology agnostic.

As in any wireless communications network, a vast majority of the performance problems in public safety networks are related to the radio access part of the network. This problem is expected to exacerbate shortly due to the worldwide trend toward LTE-based broadband public safety networks, and LTE radio access technology is an order of magnitude more complicated and more potent than traditional narrowband radio access technologies. More sophistication, more complexity, and more power in terms processing and storage capacity mean much more statistical data (big data!) to be collected and managed. Therefore, some emerging technologies and tools (such as data mining, machine learning) need to be used to evaluate the data and identify the potential performance problems.

It is also important to note that some public safety networks may employ more than one networking technology. Notably, the current trend is that the existing, already deployed narrowband technology-based networks may be augmented with LTE-based broadband networking technology (for example, P25-based technology and an LTE-based technology may be used to form a public safety network). This mixture of technologies complicates further the OSS architectures at the NML and above layers. They need to integrate the management of both narrowband and broadband technologies and the ever-increasing new and large number of services [17].

15.4 OPERATIONS SUPPORT CENTERS, POLICIES, GUIDELINES, AND PROCEDURES

In addition to OSSs and the tools discussed above, there needs to be a set of comprehensive policies, guidelines, and procedures to operate, administer, and maintain critical communications systems. These policies, guidelines, and procedures, also called Standard Operating Procedures (SOPs), establish the protocols, workflows, and procedures for the personnel of network operations as well as for the users of the system. The topics covered by these policies, guidelines, and procedures should cover all the management functional areas (i.e. configuration management, fault management, performance management, accounting management, and security management).

The list below is taken from the Department of Public Safety in Minnesota, which operates a P25-based public safety network [18].

(1) Management
  (a) Agency roles in the operational management of system
  (b) Network management
  (c) Database management
  (d) Maintenance of names and naming standards
  (e) Changing policy & standards
  (f) Security
  (g) Equipment standards
  (h) Moves, additions, and changes
  (i) Managing participation issues
  (j) Training standards
(2) Configuration and Allocation
  (a) Naming conventions
  (b) Talk-group and radio ID allocations
  (c) Fleet-mapping standards
  (d) Use of shared talk-groups

(e) Talk-group & radio user priorities

(f) Telephone interconnect

(g) Subsystem roaming

(h) Scanning

(i) Recording/Logger ports

(j) Private call

(k) Status & message transmission/warning signals

(l) Automatic Vehicle Location (AVL)/text messaging

(m) Emergency button

(n) Multi-group announcement

(3) Interoperability Guidelines

(a) MINSEF (Minnesota Statewide Emergency Frequency)

(b) Statewide fire mutual aid

(c) MIMS (Minnesota Incident Management System)

(d) Statewide EMS

(e) Recording common interagency talk-groups

(4) Guidelines for Project 25 Trunked Users

(a) Talk-group and multi-group ownership

(b) Interoperability between statewide 800 MHz system and other 800 MHz systems

(c) Statewide tactical talk-groups

(d) Interoperability between statewide 800 MHz and federal agencies

(5) Guidelines for Conventional Users

(a) Connecting to the interop system

(b) RF control stations and portables

(c) Radio to radio cross band repeaters

(6) Maintenance

(a) Agency maintenance plans

(b) Develop standards for preventive maintenance

(c) Record-keeping requirements

(d) Contact information & procedures

(e) Spare equipment

(f) Equipment configuration information

(g) Software location

(h) Notification of maintenance activities

(i) Outage responsibility/time standards/repair standards

(7) Media Policy

   (a) Media access to talk-groups

   (b) Selling radios to the media

   (c) Programming media radios

(8) Agency Billing & Cost Allocation

   (a) New users

   (b) Fees for service

   (c) Operational costs

   (d) Billing management

   (e) Insurance

(9) Compliance & Conflict Resolution

   (a) Auditing and monitoring process

   (b) Non-compliance

   (c) Appeal process

(10) Disaster Recovery Plan

   (a) Contingency procedures

   (b) Procedures/responsibility for system restoration

   (c) Levels of response

15.4.1 Centers, People, Administration

If a critical communications system is shared by several agencies, as in the case of statewide or nationwide public safety networks, a clear set of policies, guidelines, and procedures must be developed to describe the roles of each agency in the operations and administration of the network. Typically, an operation plan identifies various centers—physical locations that may contain OSSs and personnel involved in managing the network. Depending on the size of the network, there may be one or more such centers. For example, for a large statewide or nationwide public safety network, there may be several local centers and a national center. Some centers could be specialized to perform specific network management functions, and some centers may be agency specific. Note that these centers must be operational continuously (i.e. 24 hours/7 days/365 days). Also, procedures must be in place to move the center to an alternate/backup location if the conditions make it impossible to operate in the current location.

The nationwide (or statewide) center, typically called the Network Operations and Control Center (NOCC), serves as the central headquarters responsible for maintaining, administering, operating, and managing the whole network in a reliable, secure way conforming to all performance criteria established as a priority. This center is likely to be equipped with one or more OSSs to carry out its activities (not to mention that the center should have the applications, tools, policies, and people responsible for managing the network). All the centers and OSSs should be connected, and activities need to be coordinated. Some of the examples of high-level activities of the NOCC may include [18] the following:

- Hold periodic coordination meetings with the local centers to review operations and share ideas or issues in local centers that may be of interest to the local centers.
- Be available to any local center to diagnose and resolve any operational problem.
- Provide timely information to the local centers on any network issue or repair/maintenance issue.
- Monitor the performance of the entire network for normal operations.
- Monitor and manage the network databases, including regular database backups.

Typically, each local or regional center is responsible for the well-being of their portion of the network, working in close coordination and collaboration with the other centers and the NOCC. There must be clear and detailed procedures to describe these interactions, especially issues related to day-to-day operations and critical fault maintenance activities. Some of the high-level activities of a local center may include [18] the following:

- Monitor the network and its components for normal operations.
- Participate in the diagnosis of network performance problems and the development of corrective actions.
- Dispatch appropriate repair services in the event of a malfunction in the network equipment.
- Manage the database elements including subscriber IDs, talk group IDs, and various parameters.

Again, there must be policies and procedures describing how the procurement, installation, operation, and maintenance of network elements as well as network management hardware, software, and equipment are done and by which center.

To carry out the policies, guidelines, and procedures discussed above, the organization running the network needs well-qualified personnel, who are expected to have technical knowledge and experience of the system and its components. Depending on the size of the network, there may be several categories of personnel, perhaps specializing in different aspects of network operations and management. For example, some personnel may be responsible for configuration management, some may be involved in the performance management area, and yet some may be performing security management functions. Furthermore, some personnel must be involved in providing fault management functions such as performing regular maintenance and responding promptly to site repair requests.

From the governance point of view, the personnel involved in operations, administration, and management must be part of an organizational hierarchy, where the reporting structure, the roles, responsibilities, and authorities are defined in the policies and procedures.

In the case of a public safety network shared by several independent agencies, there must be a clear description of the interface among the organization of the operations support personnel of each agency and the central organization. The interface among the agencies sharing the network, as well as the interoperability of the network components and OSSs under each agency, is critically important, especially during emergency situations. The following list, taken from Reference [19], provides some examples:

- Define the complexity of your interoperability needs with a matrix of who needs to talk to whom.
- Keep procedures as simple as possible as you may not have access to your full system in an emergency.
- Protect capacity by prioritizing and limiting who will talk. Allow only critical groups to operate.
- Identify the need for unencrypted interoperability channels for external agencies.
- Use transportable networks and portable repeaters.
- Consider storing the configuration files for radio models used by your interoperability partners so you can interoperate at every level.

Another important issue here is the preparedness of the operations people, especially if several agencies are involved. As mentioned before, during emergency situations, the availability of the communications network is vitally critical. Therefore, operations personnel involved in various centers must be prepared to respond to any network failure and service outages, especially during emergency situations. Recognizing the difficulty of determining and preparing for every emergency situation, the network operations personnel, especially in a cross-functional team involving multiple agencies, must be trained and be prepared for at least the most critical and apparent emergencies (note that the training in this context is different from the training of the agents in the field).

15.4.2 Configuration Management Related Procedures

The procedures for configuration management should cover the planning, designing, installation, testing, provisioning, and deployment of the network components and the management of the physical and logical configuration data of all resources, including the components, the overall network, its services, and its subscribers. Any change in the configuration over time must be updated in the databases. The update is a mandatory activity of the configuration management functional area. However, it is equally important and necessary for the other functional areas, such as fault management and performance management, as well.

As part of the policies, guidelines, and procedures, the individual and networked configuration of each network element/component must be documented and recorded in appropriate databases.

The components that make up a critical communications system may be system level components such as network controllers, site level components such as base stations, and user level components such as portable and mobile radios. Also, almost all critical communications systems today are IP based. Therefore, they include routers, switches, and servers as well [19]. Other components to be included are the following:

- Components of the backhaul network, which may include microwave, fiber, Synchronous Optical Network (SONET), Synchronous Digital Hierarchy (SDH), Plesiochronous Digital Hierarchy (PDH), and leased line based transmission equipment
- Components of the interfaces to external networks such as Public Switched Telephone Network (PSTN) and the Internet
- Components of the IT systems providing intra-enterprise office functions, including database management

Anytime a new component is installed or an old component is taken out of service, or a component's hardware or software is upgraded, the configuration management specific procedure must be followed to configure the component and update relevant databases.

Upgrading a component, especially its software, is a configuration management activity that needs to be taken seriously since software has become a dominant part of any network component today. After any change in the configuration, the component must be thoroughly tested before rolling it out for service. Its implications, especially to the operation of other related components, must be well understood. It is also important to have a rollback plan just in case the change in the configuration causes problems when the component is rolled out.

In addition to the configuration of individual components, the configuration of the overall network and related activities must be covered by configuration management related procedures. Since the number of users (people involved in public safety) is more or less static, the configuration of the critical communication system does not change frequently. However, whenever a change occurs, such as adding a new component, replacing an existing one, or adding more transmission links, the configuration data must be managed, and there must be procedures to cover these activities.

In many cases, the initial deployment as well subsequent changes are taken care of by the vendors. The procedures should be developed to describe in detail how the operations personnel interface with the vendor; the responsibilities, procedures, workflows, management of the configuration data, etc. need to be part of these documents.

The procedures for configuration management cover the activities related to the configuration of subscriber equipment and the subscribers as well. The initial configuration information, any subsequent changes, access to specific functions, or adding or removing subscribers need to be part of these procedures. End-user devices are becoming more powerful, feature rich, more software intensive, and complicated, especially with the introduction of the LTE-capable smartphones used in critical communications systems, and therefore, managing the configuration of these devices requires carefully thought-out detailed procedures.

15.4.3 Fault Management Related Procedures

As mentioned briefly in previous sections, the area of fault management includes the prevention, detection, and correction of faults—abnormal conditions that may cause the component or the network to malfunction.

Critical communications systems, especially the ones used by public safety agencies, must be designed and built to be operational, reliable, and available continuously even under the most difficult conditions such natural and manmade disasters. Fault conditions must be proactively detected as much as possible and repaired as rapidly as possible. Managing faults can get more complicated and challenging, since a single event may generate a massive amount of faults in multiple components and transmission facilities because of the connected nature of the components. Therefore, a set of detailed, easy to understand, and easy to follow procedures are needed for the maintenance personnel to be able to analyze the fault data, identify the root cause, respond, and repair the faults as rapidly as possible.

Thanks to the highly advanced Very Large Scale Integration (VLSI) in building electronic chips and electronic boards, the need to repair the boards in the components in the field is becoming rather infrequent. The so-called field repairs involve merely the replacement of the components (such as antennas, knobs, switches) and swapping parts (such as faulty boards and faulty equipment) [19].

Also, the components must be easily accessible even during the most severe conditions so that fault management related procedures can be carried out to repair the faulty components. Sites and buildings housing the network elements and OSSs must be "hardened" with bullet-proof structures with appropriate wind load levels. Site grounding, backup power, standby equipment, fire protection, etc. are part of the preventive measures that must be described in the policies guidelines and producers [20].

In addition to the above preventive measures, which are mostly part of the fault management area, there should be clear policies and procedures describing the periodic and rigorous maintenance of the network components, sites, and OSSs. Periodic maintenance allows the operations personnel to detect and repair faults before the service is affected. Let's keep in mind, again, that the degree of strictness, the frequency of the periodic maintenance, and other preventive maintenance procedures depend on the size of the network. For smaller networks, a more reactive approach (responding only to alarms and complaints) may be taken. However, public safety networks serving public safety agencies must be proactively monitored 24/7.

The frequency of the maintenance checks also depends on the age of the network and geographical conditions such as weather (mountains and colder regions may require more frequent inspections of the sites to assess and report any weather-related potential damages). Regardless, at least an annual check on microwave systems (if used) and other backhaul transportation as well as base stations is highly recommended. For example, simulating failure conditions to test the backhaul network's switchover capability should be part of these annual maintenance procedures. Another example is the periodic testing of the receiver, transmitter, and antennas at the base station sites. Also, the backup power generators at the sites and elsewhere need to be periodically tested to make sure that they are in good working condition and have sufficient fuel [19].

Maintenance of the end-user devices (aka subscribers) must also be included in the fault management procedures. By employing various OSSs, tools, and applications, various parameters (such as transmitter and antenna) of the end device and various data (such as performance and traffic measurements) stored in the device can be monitored and modified remotely, if necessary. If a fault is detected in the user device, fault management procedures must specify the steps so that the relevant parameters may be altered, if necessary, and the related data may be downloaded for further analysis as well.

15.4.4 Performance Management Related Procedures

In general, the operations personnel involved in the performance management area are responsible for monitoring, measuring, and collecting statistical data (such as bandwidth, delay, and jitter) that can be used in assessing and reporting the behavior of the components and the network. The ultimate objective is to make sure that the network and components can be maintained at a predefined level of performance. This objective is achieved in cooperation with the planning, provisioning, and maintenance folks [4].

Unlike commercial networks, critical communications systems are more stable in the sense that modifications do not happen frequently. Leaving aside emergency situations, the network and its components operate as designed initially, requiring little optimization due to poor performance afterward.

Public safety networks are already over-engineered to handle several times more capacity than a typical busy hour. Again, depending on the type of the critical communications system and circumstances, the degree of over-engineering needs to be carefully identified since there is a tradeoff between this and the associated cost.

Obviously, for critical communications systems used by small organizations, non-safety applications, and rural agencies, the degree of over-engineering should be rather small.

Some of the performance problems may be technology related, such as poorly designed and built components, out of date components, or a poorly planned network. Updating the hardware and software of the components and increasing the capacity of transmission lines and the backhaul network may be used to boost performance.

Performance problems may also be due to the way the network is used by the users and operations personnel. Unnecessary procedures creating additional traffic load on the network, unnecessary people involved in the communications, and the system not being used in the right way are some examples in this category. For example, monitoring usage statistics on the dispatch stations to analyze the appropriateness and amount of the calls may result in some suggestions to improve the workload. A comprehensive and periodic training effort for the agents and operations personnel to become "fluent" on the network, its components, as well as on the procedures and policies may minimize some of the performance problems due to people.

Finally, due to the inclusion of emerging technologies (packetization, softwarization, virtualization, etc.) in the network, performance management has become very complicated. Degradation of performance may be detected by proactively monitoring delay, packet loss, and congestion in the network even when there is no physical component, equipment, or facility failure. The difficulty is in the process of pinpointing the actual component(s) causing the problem and requiring a solution to alleviate the problem.

15.4.5 Accounting Management Related Procedures

The functional area of accounting management is responsible for measuring the use of services in the network and determining the amount of money the user needs to pay for this usage. Therefore, typical accounting management functions include usage measurement, pricing, and collection of payments. Pricing typically includes the use of a tariff to calculate the amount of money that must be collected. The type of measurements collected for accounting management must be limited to the data related to the cost of the devices and services (other management functional areas collect their data).

Accounting management, especially the pricing and payments by the subscriber may not make sense for critical communications systems owned and used by the employees of a single organization. However, depending on how the network was built, operated, and shared, it may require pricing functions. For example, if the network is built and maintained by a vendor based on a financial agreement that requires that usage be measured, the agencies are billed for usage for a certain period. Also, if the network is used by several agencies, or the network is a private–public partnership type, then this functional area becomes essential. The cost of deploying and operating the network must be shared by the all partners based on prearranged negotiated rules, which may require that the users and agencies get billed based on their usage of the network. Usage measurement may be monitored and collected at the user level, device level, and at the agency level [21].

The usage measurement collected as part of accounting management may also be useful for the operations of other functional areas, such as fault, performance, and security management. For example, through these measurements, an agency can extract necessary information such as the number of devices in use, the status of upgrades, and the usage patterns of agents. Any abnormal amount of measurements—below or above certain thresholds—may be the cause for flags that may trigger security, performance, or fault management actions.

15.4.6 Security Management Related Procedures

The policies, guidelines, and procedures in this category are dedicated to ensuring that the critical communication system is secure and helps deliver uninterrupted services.

The systems and the networks are typically embedded and integrate a set of extensive security-specific features, such as encrypted transmission, authentication, and authorization. However, a set of policies, guidelines, and procedures need to be developed to make sure any threats to the security of critical communications systems can be prevented. In the case of an intrusion, the policies and procedures should describe how the intrusion can be detected, isolated, and contained, and if there is a disruption, how the service is restored.

These security management related guidelines and procedures should include the security of the sites, which include base stations and antenna towers as well as the buildings that house other network elements, OSSs, and operations and maintenance personnel. Fencing with barbed razor wires, security cameras, and alarms at the gates and doors are some of the typical measures taken [20].

As a preventive measure, for example, the public safety network should be isolated from the rest of the network, which the agency may be using for its routine office procedures. Also, a regular periodic review of security management related policies, guidelines, practices, and procedures should be performed as part of preventive measures. The use of encryption must be enforced; frequent change of the encryption key should be part of these procedures [22].

Security-related information collected by the components of the network as well as by the OSS should regularly be analyzed to determine any unusual activity, threats, unnoticed external or internal breaches, etc. in the agency (it is possible that disgruntled internal personnel may be the culprit for threats).

Some common sense practices that are used in many agencies include the logging of events and features to keep track of who accessed what, enforcing the use of strong passwords, configuration of various protocols with robust security features, and exponential backoffs.

Identity management has become a critical part of security management. Identification of individuals, devices, and the networks with a unique set of predefined and pre-assigned attributes is essential to authenticate and authorize access to services and resources. The network/service must be able to identify and authenticates users and devices. In return, users and devices must be able to identify the network as well to make sure the network is the right network, not a bogus one.

Identity management requires even more strict rules and procedures in critical communication systems using LTE technology. Much higher data rates, multimedia and content-rich information, involvement of potentially several partner organizations using the same network, and a variety of new services will make identity management more difficult to manage. New capabilities such as wearables and the use of biometric technology based authentication for end-user devices will be needed [23].

15.5 SUMMARY AND CONCLUSIONS

This chapter focused on network management related topics for critical communications systems. The chapter provided a detailed discussion on the operation, administration, and maintenance of critical communications systems. After a brief introduction of the generic network and service management concepts, the chapter discussed how these generic concepts applied to critical communications systems.

For large critical communications systems, a separate "network and service management" framework, called TMN, may exist to handle the maintenance, operations, administration, and provisioning of the actual network. The TMN framework includes a set of OSSs that are connected to each other as well as to the components of the critical communications system. The framework also includes applications, plans, policies, procedures, and people to provide a variety of functions under the five management functional areas—fault management, performance management, configuration management, security management, and accounting management [4].

Developing a master plan called the OA&M plan, or just the operations plan, is necessary to identify and describe the OSSs, management centers, tools, applications, policies, and procedures needed to run the network. The next step is to acquire or build the systems that will support the operations and develop detailed policies, guidelines, and procedures to be followed to handle various related activities to manage the network.

Network and service management are usually and unfortunately an afterthought in commercial networks. Network management is especially important for public safety networks, which demand the utmost care, continuous monitoring, and taking necessary steps to intervene, if necessary, as soon as possible. There must be capabilities to detect and repair any fault and anomalies with minimum disruption of the communications service. Furthermore, with LTE-based public safety networks, running the network will become more complex and will require a significant focus on the management of networks and services at the very first effort to plan and deploy such networks.

REFERENCES

1. "M.3010: Principles for a telecommunications management network." ITU-T Recommendation, 2000.
2. S. Aidarous and T. Plevyak, Eds., *Telecommunications Network Management into the 21st Century.* IEEE Press, 1994.
3. T. Plevyak and V. Sahin, Eds., *Next Generation Telecommunications Networks, Services, and Management.* Wiley/IEEE Press, 2010.
4. "M.3400: TMN management functions." ITU-T Recommendation, 2000.
5. A. Clemm, *Network Management Fundamentals.* Cisco Press, 2007.
6. Montana Statewide Interoperability Governing Board, "Public safety land mobile radio system operations and maintenance plan," Nov. 2014. Available: https://sitsd.mt.gov/ LinkClick.aspx?fileticket=ivqLfshmdGc%3D&portalid=77.
7. H. M. GmbH, "Network management system (NMS)." [Online]. Available: https://www .hytera-mobilfunk.com/en/technologies/tetra/network-management-system-nms/. [Accessed: Mar. 31, 2018].
8. Airbus Co., "Element Monitor, ElMo." [Online]. Available: https://www.secureland communications.com/element-monitoring. [Accessed: Apr. 1, 2018].
9. J. Dunlop, D. Girma, and J. Irvine, *Digital Mobile Communications and the TETRA System.* John Wiley & Sons, 2013.
10. Kalibre, "Implementing a Manager of Managers for effective fault management of public safety radio networks," White Paper, 2012.
11. Teltronic, "eNEBULA TETRA infrastructure." [Online]. Available: http://www.teltronic .es/en/products/tetra/enebula-tetra-infrastructure/. [Accessed: Apr. 1, 2018].
12. "Network Management Systems." [Online]. Available: https://www.hytera-mobilfunk .com/en/technologies/tetra/network-management-system-nms/. [Accessed: Jun. 25, 2016].
13. D. Mauro and K. Schmidt, *Essential SNMP*, 2nd ed. O'Reilly Media, 2005.
14. "TETRA compliant alarm monitoring software." [Online]. Available: https://www. copybook.com/companies/innovative-business-software/alarm-monitoring-software-and- incident-management-solutions-gallery/tetra-compliant-alarm-monitoring-software. [Accessed: Apr. 3, 2018].
15. City of Fort Worth, "P25 system management access policy for external agencies with infrastructure," Aug. 1, 2014. [Online]. Available: http://fortworthtexas.gov/uploaded Files/IT_Solutions/Radios/ITS-RADIO-010_SysMgmt_External_Agencies.pdf.
16. Motorola Solutions, "Integrated terminal management." [Online]. Available: https://www .motorolasolutions.com/content/dam/msi/docs/business/product_lines/dimetra_tetra/ terminals/tetra_integrated_terminal_management_system/_documents/itm_specsheet_ 08_10_upload.pdf. [Accessed: Sep. 15, 2018].
17. Alcatel-Lucent, "Operating and managing TETRA networks," 2010. [Online]. Available: http://enterprise.alcatel-lucent.com/assets/documents/SBG5677100105_OSS_Tetra_EN_ Brochure.pdf. [Accessed: Sep. 15, 2018].
18. Department of Public Safety, Minnesota, "P25 network management standards," 2017. [Online]. Available: https://dps.mn.gov/divisions/ecn/programs/armer/Documents/ Network%20Standards.pdf.

19. Tait Communications, "P25 best practice presented by Tait Communications," P25 Best Practice, Jan. 23, 2014. [Online]. Available: http://www.p25bestpractice.com. [Accessed: Jan. 10, 2017].
20. E. Wibbens, "Public safety site hardening," Harris Corporation, White Paper, 2014.
21. T. Hengeveld, "Public safety entity control and monitoring requirements for the Nationwide Public Safety Broadband Network," NPSTC, LC14-07-21, Oct. 2015.
22. S. Jacobs, *Security Management of Next Generation Telecommunications Networks and Services*. John Wiley & Sons, 2013.
23. N. Hastings and J. Franklin, "Consideration for identity management in public safety mobile networks," National Institute of Standards and Technology, NIST IR 8014, Mar. 2015.

16

SUMMARY AND CONCLUSIONS

This chapter provides a summary of the major points addressed in this book. The chapter also discusses challenges that decision-makers may face with the deployment of critical communications systems.

16.1 MAJOR POINTS AND OBSERVATIONS

Today, the analog public safety radio technologies used for critical communications systems have mostly been replaced by narrowband, all-digital, and voice and data technologies led by Project 25, Terrestrial Trunked Radio (TETRA), and Digital Mobile Radio (DMR) standards.

TETRA has been the choice of public safety agencies mainly in Europe and Project 25 technologies primarily in North America, but both have worldwide deployments as well. DMR-based systems have also been chosen in some regions around the world due to their simplicity, backward compatibility, and cost-effectiveness.

Fundamentals of Public Safety Networks and Critical Communications Systems: Technologies, Deployment, and Management, First Edition. Mehmet Ulema.
© 2019 by The Institute of Electrical and Electronics Engineers, Inc. Published 2019 by John Wiley & Sons, Inc.

Project 25 and TETRA technologies are mature, widely used, tested, reliable, and feature-rich in voice-based applications.

A significant majority of narrowband technologies have been used only in individual agency based systems. In other words, there are not too many nationwide systems that use narrowband technologies. This implies that they are not really tested and tried enough, so we do not know how successfully they will scale up and behave in a nationwide system.

The public safety systems based on narrowband technology components are more expensive than the Long Term Evolution (LTE) based systems, since the target market is relatively small (limited to only critical communications applications), compared to the LTE technology used in commercial mobile applications (the total number of worldwide users of narrowband technologies is about 40 million, compared to multibillion commercial mobile users).

Narrowband technologies are limited in providing data services. The data rates, which range from 9.6 Kbps to 36 Kbps, are relatively low compared to what modern data applications require today. The demand for data-intensive applications by public safety agencies is increasing. Laptops, notebooks, and tablets (typically mounted in police vehicles) and handheld devices for ticketing applications are becoming commonplace. More recently, public safety organizations have begun using smartphones and tablets.

The socioeconomic benefits of using broadband data are mind-boggling; as discussed in Chapter 13, the annual consolidated socioeconomic benefits computed for European Union (EU) countries with a total combined population of 500 million would be approximately 34 billion Euros. The opportunity cost is relatively small. For the European scenario mentioned above, selling the spectrum at auction would obtain a one-off economic gain for the government, totaling 3.7 billion Euros. The last two points indicate that the benefits are several times greater than the opportunity cost, suggesting that there is no doubt about the economic feasibility of implementing broadband-based public safety networks.

Because of the technical and economic reasons as well as the users' needs discussed above, it is abundantly clear that LTE is an ideal technology to build and deploy a critical communications system. LTE is especially suitable for a nationwide broadband public safety network for all the public safety agencies to use in a cooperative and interoperable manner. LTE is a proven and tested technology for commercial use and for nationwide broadband networks. It handles broadband data applications an order of magnitude better than narrowband systems do. The scale of economy is just outstanding. For the first time, in the history of commercial mobile communications, LTE has emerged as the single worldwide standard used all around the world, involving many vendors, operators, app developers, technical expertise, etc.

The standards organizations and stakeholders involved in developing narrowband technologies have indicated publicly that their future strategy is to evolve into using LTE-based solutions for public safety systems. A growing number of countries, including the USA, have chosen LTE for their public safety networks already.

The FirstNet in the USA has been established with a mandate to carry out the implementation of a nationwide LTE-based broadband safety network.

Although the LTE standards include features to support voice communications and applications, as of writing this book, many commercial LTE operators use LTE only for data communications (some operators around the world are already offering voice services over LTE, aka Voice over LTE [VoLTE]). More importantly, several public safety specific voice applications (e.g. Push-To-Talk [PTT]) are just in the process of being incorporated into standards and vendor equipment. They may take a while to make it into the baseline products that are widely available.

Although many new critical communications systems are expected to be based on LTE technology, the older technologies such as Project 25 and TETRA will not go away. For a variety of reasons, the new critical communication system may be based on a narrowband technology. Anyhow, in the future critical communications landscape, we will see some systems based on narrowband technologies and some others based on broadband technologies, perhaps interoperating seamlessly to provide the broadest coverage possible as well as to be used by as many agencies as possible.

16.2 CHALLENGES IN DEPLOYING CRITICAL COMMUNICATIONS SYSTEMS

Persuading the public safety community and other stakeholders to accept a new (or enhanced) public safety system may not be an easy task. There will be some challenges to overcome. While some of these challenges are real concerns, others are typically more of "reaction to change" type concerns, which can be overcome by reaching out and providing training, education, and publicity.

It is possible that the new network and underlying technology may not work as expected. The legacy services provided by the existing technology may not work as satisfactorily in the new system. For example, the prioritization feature promised by LTE technology may not work correctly, resulting in the users with less priority not being pre-emptible. The LTE technology with the shared spectrum feature may not work well, creating more congestion. These are all valid concerns. However, LTE technology has been around and has been used in many commercial applications, namely 4G cellular mobile systems, around the world. Therefore, there is an ongoing worldwide effort to improve and enhance LTE technology through standardization as well as through some vendor-defined/implemented solutions. Additionally, typically new systems are designed with flexibility to adjust themselves and, if necessary, default to primary operating modes, consistent with legacy system operations.

Another concern has to do with the frequency spectrum. In the case of the deployment of a region-wide or nationwide public safety network based on LTE technology, the new system may require some changes in regulatory policies, especially in spectrum allocation. Therefore, it is essential that regulatory hurdles be taken care of before the project is launched. Also, the political situation in the country or the state and resistance by those with vested interests in maintaining the status quo (vendors, suppliers, service providers, etc.) may also impact the project.

It should be recognized that anytime a new technology/system is introduced, the early users (aka early adaptors) face higher costs and lower benefits. This may discourage other users from joining. However, in the case of LTE technology, this concern should not be a big problem to overcome, since LTE has been in use in many commercial networks, and therefore there will not be early adapter high cost. A related challenge could be that the price of the new, LTE-based public safety network could be higher than initially estimated. Since the LTE market is spread worldwide, is extensive, and is highly competitive, the equipment and deployment cost will not fluctuate upward. Also, as discussed in Chapter 13, there are many financing and economic options to consider to alleviate this problem. For example, there is an opportunity to generate some revenue by letting commercial operators use the excess capacity.

Another challenge to overcome is the perception that the new systems based on LTE will not be as reliable as the existing public safety systems. It is true that the new communications technologies are more software intensive and virtualized and therefore introduce real challenges in the reliability area. Luckily, LTE technology has shown that networks can be designed and built with similar or even better reliability requirements than those of public safety networks today.

There are many stakeholders in the public safety ecosystem—public safety organizations, institutions, operators, regulatory agencies, equipment and device suppliers, volunteers, legislators, and beneficiaries of public safety activities. They all have different, sometimes competing, interests and positions, but let's keep in mind that the issue of public safety is the main reason they all should be together in this. Leadership is needed to coordinate all these and to take advantage of the strengths of each stakeholder to carry out their work. Overcoming the resistance of those with vested interests and educating them is an essential challenge. However, an intensive outreach and education campaign focusing on the critical importance and necessity of the public safety network with new technologies, additional capabilities, and with new opportunities should help overcome this challenge as well.

A

PROJECT 25 DOCUMENTS

A.1 TELECOMMUNICATION INDUSTRY ASSOCIATION (TIA) DOCUMENTS

Note 1: Only the active documents are listed.

TIA Document #	Official Document Title	Published
TSB-102-B	Project 25 TIA-102 Documentation Suite Overview	Jun. 2012
TIA-102.AAAB-A	Project 25 Digital Land Mobile Radio – Security Services Overview	Jan. 2005
TIA-102.AAAB-A-1	Project 25 Digital Land Mobile Radio – Security Services Overview - Addendum Key Management Architecture	Sep. 2014
TIA-102.AAAD-B	Project 25 Digital Land Mobile Radio Block Encryption Protocol	Dec. 2015

Fundamentals of Public Safety Networks and Critical Communications Systems: Technologies, Deployment, and Management, First Edition. Mehmet Ulema.
© 2019 by The Institute of Electrical and Electronics Engineers, Inc. Published 2019 by John Wiley & Sons, Inc.

TIA Document #	Official Document Title	Published
TIA-102.AABA-B	Project 25 Trunking Overview Digital Radio Technical Standards	Apr. 2011
TIA-102.AABB-B	Project 25 Trunking Control Channel Formats	July. 2011
TIA-102.AABC-D	Project 25 Trunking Control Channel Messages	Apr. 2015
TIA-102.AABD-B	Project 25 Trunking Procedures	Nov. 2014
TIA-102.AABF-D	Project 25 Link Control Word Formats and Messages	Apr. 2015
TIA-102.AABG	Project 25 Conventional Control Messages	Apr. 2009
TIA-102.AABH	Project 25 Dynamic Regrouping Messages and Procedures	Nov. 2014
TIA/EIA-102.AACA-A	Project 25 Digital Radio Over-the-Air Rekeying (OTAR) Protocol	Sep. 2014
TIA-102.AACC-A	Conformance Tests for Project 25 Over-The-Air Rekeying (OTAR) Protocol	Aug. 2006
TIA-102.AACD-A	Project 25 Digital Land Mobile Radio - Key Fill Device (KFD) Interface Protocol	Sep. 2014
TIA-102.AACE-A	Project 25 Digital Land Mobile Radio Link Layer Authentication	Apr. 2011
TIA-102.BAAA-A	Project 25 FDMA - Common Air Interface	Sep. 2003
TIA-102.BAAB-B	Project 25 Common Air Interface Conformance Test	Mar. 2005
TIA-102.BAAC-C	Project 25 Common Air Interface Reserved Values	Apr. 2011
TIA-102.BAAD-B	Project 25 Conventional Procedures	Aug. 2015
TIA-102.BABA-A	Project 25 Vocoder Description	Feb. 2014
TIA/EIA-102.BABB	Project 25 Vocoder Mean Opinion Score Conformance Test	May. 1999
TIA/EIA-102.BABC	Project 25 Vocoder Reference Test	Apr. 1999
TIA-102.BABG	Project 25 Enhanced Vocoder Methods of Measurement for Performance	Mar. 2010
TIA-102.BACA-B	Project 25 Inter-RF Subsystem Interface Messages and Procedures for Voice and Mobility Management Services	Nov. 2012
TIA-102.BACA-B-1	Project 25 Inter-RF Subsystem Interface Messages and Procedures for Voice, Mobility Management, and RFSS Capability Polling Services - Addendum 1 - Group Emergency Behaviors	Jul. 2013

TIA Document #	Official Document Title	Published
TSB-102.BACC-B	Project 25 Inter-RF Subsystem Interface Overview	Nov. 2011
TIA-102.BACD-B	Project 25 Inter-RF Subsystem Interface (ISSI) Messages and Procedures for Supplementary Data	Jul. 2011
TIA-102.BACE	Project 25 Inter-RF Subsystem Interface (ISSI) Messages and Procedures for Conventional Operations	Jun. 2008
TIA-102.BACF	Project 25 Inter-RF Subsystem Interface (ISSI) Messages and Procedures for Packet Data Services	Oct. 2009
TIA/EIA-102.BADA-A	Project 25 Telephone Interconnect Requirements and Definitions (Voice Services)	June. 2012
TIA/EIA-102.BAEA-C	Project 25 Data Overview	Dec. 2015
TIA-102.BAEB-B	Project 25 IP Data Bearer Service Specification	Sep. 2014
TIA-102.BAED	Project 25 Packet Data Logical Link Control Procedures	Oct. 2013
TIA-102.BAEE-C	Project 25 Radio Management Controls	Dec. 2015
TIA-102.BAEF	Project 25 Packet Data Host Network Interface	Nov. 2013
TIA-102.BAEG	Project 25 Mobile Data Peripheral Interface	May. 2013
TIA-102.BAEJ	Project 25 Conventional Management Service Specification for Packet Data	Oct. 2013
TSB-102.BAFA-A	Project 25 Network Management Interface Overview	Jul. 2099
TSB-102.BAGA	Project 25 Console Subsystem Interface Overview	Feb. 2008
TIA-102.BAHA	Project 25 Fixed Station Interface Messages and Procedures	Jun. 2006
TSB-102.BAJA-A	Project 25 Location Services Overview	Feb. 2010
TIA-102.BAJB. 20A	Project 25 Tier 1 Location Services	Nov. 2014
TIA-102.BAJC-A	Project 25 Tier 2 Location Services Specifications	Apr. 2015
TIA-102.BAJD	Project 25 TCP/UDP Port Number Assignments	Oct. 2010
TIA-102.BAKA	Project 25 KMF to KMF Interface	Apr. 2012
TSB-102.BBAA	Project 25 Two-slot TDMA Overview	Mar. 2010
TIA-102.BBAB	Project 25 Phase 2 Two-Slot Time Division Multiple Access Physical Layer Protocol Specification	Jul. 2009

TIA Document #	Official Document Title	Published
TIA-102.BBAC	Project 25 Phase 2 Two-slot TDMA Media Access Control Layer Description	Dec. 2010
TIA-102.BBAC-1	Project 25 Phase 2 Two-slot TDMA Media Access Control Layer Description Addendum 1	Feb. 2013
TIA-102.BCAD	Project 25 Phase 2 Two-Slot Time Division Multiple Access Trunked Voice Services Common Air Interface Conformance Specification	Sep. 2011
TIA-102.BCAE	Project 25 Phase 2 Two-Slot Time Division Multiple Access Trunked Voice Services Message and Procedures Conformance Specification	Jul. 2011
TIA-102.BCAF	Project 25 Trunked TDMA Voice Channel Conformance Profiles	Aug. 2012
TIA-102.CAAA-D	P25 Digital C4FM/CQPSK Transceiver Measurement Methods	Apr. 2013
TIA-102.CAAB-D	Project 25 Land Mobile Radio Transceiver Performance Recommendations - Digital Radio Technology C4FM/CQPSK Modulation	Feb. 2013
TSB-102.CAAC	Project 25 Mobile Radio Push-to-Talk and Audio Interface - Definitions and Methods of Measurement	Sep. 2002
TIA-102.CABA	Project 25 Interoperability Test Procedures Conventional Voice Equipment	Oct. 2010
TIA-102.CABB-A	Project 25 Interoperability Testing for OTAR Messages and Procedures	Nov. 2015
TIA-102.CABC-B	Project 25 Interoperability Testing for Voice Operation in Trunked Systems	Nov. 2010
TIA-102.CABC-B-1	Project 25 Interoperability Testing for Voice Operation in Trunked Systems Addendum - TDMA Mode	Jul. 2011
TIA-102.CACA	Project 25 Inter RF Subsystem Interface (ISSI) Measurement Methods for Voice Services	Apr. 2007
TIA-102.CACA-1	Project 25 Inter RF Subsystem Interface (ISSI) Measurement Methods for Voice Services - Addendum 1 Trunked Console ISSI	Dec. 2008
TIA-102.CACB	Project 25 Inter-RF Subsystem Interface (ISSI) Performance Recommendations for Voice Services	Apr. 2007
TIA-102.CACB-1	Project 25 Inter-RF Subsystem Interface (ISSI) Performance Recommendations for Voice Services - Addendum 1 Trunked Console ISSI	Dec. 2008

TIA Document #	Official Document Title	Published
TIA-102.CACC	Project 25 Inter-RF Subsystem Interface Conformance Test Procedures	Jan. 2009
TIA-102.CACC-1	Project 25 Inter-RF Subsystem Interface Conformance Test procedures Addendum 1 - Supplementary Data	Aug. 2011
TIA-102.CACD-C	Project 25 Inter-RF Subsystem Interface - Interoperability Test Procedures for Trunked Systems Involving the ISSI	Nov. 2014
TIA-102.CADA	Project 25 Fixed Station Interface Conformance Test Procedure	Apr. 2007
TIA-102.CAEA	Project 25 Conformance Profile Level One - Basic Conventional Operation	Apr. 2009
TIA-102.CAEB	Project 25 Conformance Profile Level Two - Advanced Conventional Operation	Dec. 2009
TIA-102.CAEC	Project 25 Conformance Profile Level Three - Basic Trunked Operation	Dec. 2011
TIA-102.CAED	Project 25 Conformance Profiles for Advanced Trunked Operations	Nov. 2013
TSB-102.CBAA	Project 25 Compliance Assessment Overview	Oct. 2010
TSB-102.CBAB	Project 25 Compliance Assessment Program Supplier Declaration of Compliance Template	Jun. 2009
TSB-102.CBAC	Project 25 Compliance Assessment Summary Test Report Guidelines - Transceiver Performance	Apr. 2009
TSB-102.CBAF	Project 25 Compliance Assessment Summary Test Report Guidelines - Trunking Interoperability	Apr. 2009
TSB-102.CBBA-A	Project 25 Compliance Assessment Tests - Transceiver Performance - Conventional Mode Subscriber	Jun. 2015
TSB-102.CBBC-A	Project 25 Compliance Assessment Tests - Transceiver Performance - Conventional Mode Fixed Station	Jun. 2015
TSB-102.CBBE	Project 25 Recommended Compliance Assessment Tests - Conventional Operation	Sep. 2011
TSB-102.CBBF-A	Project 25 Recommended Compliance Assessment Tests - Transceiver Performance - Trunking Mode Subscriber	Jun. 2015
TSB-102.CBBH-A	Project 25 Recommended Compliance Assessment Tests - Performance - Trunked Mode Fixed Station Transceiver and Related Infrastructure	May. 2015

TIA Document #	Official Document Title	Published
TSB-102.CBBJ-C	Project 25 Recommended Compliance Assessment Tests - Trunking Interoperability	Jun. 2015
TSB-102.CBBK-A	Project 25 Recommended Compliance Assessment Tests - Trunking ISSI	Mar. 2010
TSB-102.CBBL-A	Project 25 Recommended Compliance Assessment Tests for TDMA Trunking Voice Channel Air Interface	Jun. 2015
TIA-102.CCAA-A	Project 25 Two-Slot Time Division Multiple Access Transceiver Measurement Methods	May. 2014
TIA-102.CCAB-A	Project 25 Two-Slot Time Division Multiple Access Transceiver Performance Recommendations	Mar. 2014

B

TETRA DOCUMENTS BY ETSI

Source: "ETSI TETRA Standards." [Online]. Available: http://www.oborne consulting.co.uk/TETRA_Standards/ETSI_TETRA_Standards.htm#. [Accessed: 04-Mar-2017].

Note 1: European Telecommunication Standards Institute (ETSI) TETRA & Critical Communications Evolution standards are produced by ETSI Technical Committee TETRA and Critical Communications Evolution (TC TCCE, formally TC TETRA). Various types of standards documents are produced:

- All new documents comply with the "new regime" which includes ENs, ESs, EGs, TSs, and TRs
- Previously, "old regime" documents (of which some are still valid) included ETSs, ETRs, and TBRs

Note 2: Only the currently active documents are listed. Historical documents which are not maintained and deleted and withdrawn documents are not listed.

Fundamentals of Public Safety Networks and Critical Communications Systems: Technologies, Deployment, and Management,
First Edition. Mehmet Ulema.
© 2019 by The Institute of Electrical and Electronics Engineers, Inc. Published 2019 by John Wiley & Sons, Inc.

Note 3: Edition and Version numbers and publication dates are indicated with the letter "V," the year and month, respectively.

B.1 EUROPEAN TELECOMMUNICATION STANDARDS (ETSs) AND ETSI EUROPEAN STANDARDS (ENs)

	TETRA Voice + Data (V+D)	
EN 300 392-1	Voice + Data General Network Design	V1.4.1; 2009-01
EN 300 392-2	Voice + Data Air Interface (AI)	V3.8.1; 2016-08
EN 300 392-3	**Voice + Data Interworking - basic operation**	
EN 300 392-3-1	Interworking at the Inter-System Interface (ISI); General Design	V1.4.1; 2015-12
EN 300 392-3-2	Interworking at the Inter-System Interface (ISI); Additional Network Feature Individual Call (ANF-ISIIC)	V1.4.1; 2010-08
EN 300 392-3-3	Interworking at the Inter-System Interface (ISI); Additional Network Feature Group Call (ANF-ISIGC)	V1.3.1; 2011-11
EN 300 392-3-4	Interworking at the Inter-System Interface (ISI); Additional Network Feature Short Data Service (ANF-ISISDS)	V1.3.1; 2010-08
EN 300 392-3-5	Interworking at the Inter-System Interface (ISI); Additional Network Feature for Mobility Management (ANF-ISIMM)	V1.5.1; 2016-06
392-3-6, -7, -8	Note - see TS 100 392-3-6, -7 & -8 for subsequent parts of ISI interworking	
ETS 300 392-4	**Voice + Data Gateways - basic operation**	
ETS 300 392-4-1	V+D Gateways - basic operation: PSTN	Edition 1; Jan 99
ETS 300 392-4-2	V+D Gateways - basic operation: ISDN Gateway	Edition 1; Sep 00
EN 300 392-5	Peripheral Equipment Interface	V2.5.1; 2016-10
EN 300 392-7	Voice + Data Security	V3.4.1; 2017-01
EN 300 392-9	General Requirements for Supplementary Services	V1.5.1; 2012-04
EN 300 392-10	**V+D Supplementary Services (SS) Stage 1**	
EN 300 392-10-1	SS Stage 1: Call identification	V1.3.1; 2004-01
ETS 300 392-10-2	SS Stage 1: Call report (CR)	Edition 2; Aug 00

ETS 300 392-10-3	SS Stage 1: Talking party identification (TPI)	Edition 2; Jul 99
EN 300 392-10-4	SS Stage 1: Call Forwarding (CF)	V1.3.1; 2003-09
ETS 300 392-10-5	SS Stage 1: List Search Call (LSC)	Edition 2; Aug 00
EN 300 392-10-6	SS Stage 1: Call authorized by dispatcher (CAD)	V1.4.1; 2006-08
ETS 300 392-10-7	SS Stage 1: Short number addressing	Edition 2; Sep 99
EN 300 392-10-8	SS Stage 1: Area selection (AS)	V1.2.1; 2004-02
ETS 300 392-10-9	SS stage 1: Access Priority	Edition 2; Dec 98
EN 300 392-10-10	SS Stage 1: Priority call (PC)	V1.2.1; 2002-05
EN 300 392-10-11	SS Stage 1: Call waiting (CW)	V1.3.1; 2004-01
EN 300 392-10-12	SS Stage 1: Call hold (CH)	V1.3.1; 2004-02
ETS 300 392-10-13	SS Stage 1: Call completion to busy subscriber	Edition 2; Sep 99
EN 300 392-10-14	SS Stage 1: Late entry (LE)	V1.2.1; 2002-09
EN 300 392-10-16	SS Stage 1: Pre-emptive priority call (PPC)	V1.3.1; 2006-08
EN 300 392-10-17	SS Stage 1: Include call (IC)	V1.2.1; 2002-05
EN 300 392-10-18	SS Stage 1: Barring of outgoing calls (BOC)	V1.3.1; 2003-10
EN 300 392-10-19	SS Stage 1: Barring of incoming calls (BIC)	V1.2.1; 2002-09
ETS 300 392-10-20	SS Stage 1: Discreet listening (DL)	Edition 2; May 99
EN 300 392-10-21	SS Stage 1: Ambience listening (AL)	V1.2.1; 2003-09
EN 300 392-10-22	SS Stage 1: Dynamic group number assignment (DGNA)	V1.2.1; 2002-01
ETS 300 392-10-23	SS Stage 1: Call completion on no reply	Edition 2; Sep 99
ETS 300 392-10-24	SS Stage 1: Call retention (CRT)	Edition 2; Apr 00
EN 300 392-11	**V+D Supplementary Services (SS) Stage 2**	
EN 300 392-11-1	SS Stage 2: Call identification (CI)	V1.2.1; 2004-01
ETS 300 392-11-2	SS Stage 2: Call report (CR)	Edition 1; Sep 00
ETS 300 392-11-3	SS Stage 2: Talking party identification	Edition 1; Jul 99
EN 300 392-11-4	SS Stage 2: Call Forwarding (CF)	V1.1.1; 2003-07
ETS 300 392-11-5	SS Stage 2: List Search Call (LSC)	Edition 1; Sep 00
EN 300 392-11-6	SS Stage 2: Call authorized by dispatcher (CAD)	V1.2.1; 2004-01

(Continued)

ETS 300 392-11-7	SS Stage 2: Short number addressing (SNA)	Edition 1; Apr 00
EN 300 392-11-8	SS Stage 2: Area selection (AS)	V1.1.1; 2000-12
ETS 300 392-11-9	SS Stage 2; Access Priority (AP)	Edition 1; Oct 98
EN 300 392-11-10	SS Stage 2: Priority call (PC)	V1.1.1; 2001-05
ETS 300 392-11-11	SS Stage 2: Call waiting (CW)	Edition 1; Sep 00
EN 300 392-11-12	SS Stage 2: Call hold (CH)	V1.1.2; 2003-05
ETS 300 392-11-13	SS Stage 2: Call completion to busy subscriber (CCBS)	Edition 1; Mar 00
EN 300 392-11-14	SS Stage 2: Late entry (LE)	V1.1.1; 2002-07
ETS 300 392-11-15	SS Stage 2: Transfer of control (withdrawn)	n/a
EN 300 392-11-16	SS Stage 2: Pre-emptive priority call (PPC)	V1.2.1; 2004-09
EN 300 392-11-17	SS Stage 2: Include call (IC)	V1.1.2; 2002-01
EN 300 392-11-18	SS Stage 2: Barring of outgoing calls (BOC)	V1.1.1; 2001-08
EN 300 392-11-19	SS Stage 2: Barring of incoming calls (BIC)	V1.1.1; 2001-08
ETS 300 392-11-20	SS Stage 2: Discreet listening (DL)	Edition 1; Aug 99
EN 300 392-11-21	SS Stage 2: Ambience listening	V1.1.1; 2003-04
ETS 300 392-11-22	SS Stage 2: Dynamic group number assignment (DGNA)	Edition 1; Apr 00
ETS 300 392-11-23	SS Stage 2: Call completion on no reply (CCNR)	Edition 1; Apr 00
ETS 300 392-11-24	SS Stage 2: Call retention (CRT)	Edition 1; Sep 00
ETS 300 392-11-25	SS Stage 2: Advice of charge (withdrawn)	n/a
EN 300 392-12	**V+D Supplementary Services (SS) Stage 3**	
EN 300 392-12-1	SS Stage 3: Call identification (CI)	V1.2.2; 2007-08
ETS 300 392-12-2	SS Stage 3: Call report (CR)	Edition 1; Sep 00
EN 300 392-12-3	SS Stage 3: Talking party identification (TPI)	V1.3.1; 2006-04
EN 300 392-12-4	SS Stage 3: Call Forwarding	V1.4.1; 2016-07
ETS 300 392-12-5	SS Stage 3: List Search Call (LSC)	Edition 1; Sep 00
EN 300 392-12-6	SS Stage 3; Call Authorized by Dispatcher (CAD)	V1.3.1; 2006-02

ETS 300 392-12-7	SS Stage 3: Short number addressing (SNA)	Edition 1; Apr 00
EN 300 392-12-8	SS Stage 3: Area selection (AS)	V1.2.1; 2010-07
ETS 300 392-12-9	SS Stage 3; Access Priority	Edition 1; Oct 98
EN 300 392-12-10	SS Stage 3: Priority call (PC)	V1.2.1; 2004-02
ETS 300 392-12-11	SS Stage 3: Call waiting (CW)	Edition 1; Sep 00
EN 300 392-12-12	SS Stage 3: Call hold (CH)	V1.1.2; 2003-05
EN 300 392-12-13	SS Stage 3: Call completion to busy subscriber (CCBS)	V1.2.1; 2012-03
EN 300 392-12-14	SS Stage 3: Late entry (LE)	V1.2.1; 2012-03
EN 300 392-12-16	SS Stage 3: Pre-emptive priority call (PPC)	V1.2.1; 2004-09
EN 300 392-12-17	SS Stage 3: Include call (IC)	V1.1.2; 2002-01
EN 300 392-12-18	SS Stage 3: Barring of outgoing calls (BOC)	V1.1.1; 2001-08
EN 300 392-12-19	SS Stage 3: Barring of incoming calls (BIC)	V1.1.1; 2001-08
EN 300 392-12-20	SS Stage 3: Discreet listening	V1.2.1; 2012-04
EN 300 392-12-21	SS Stage 3: Ambience listening (AL)	V1.5.1; 2012-04
EN 300 392-12-22	SS Stage 3: Dynamic group number assignment (DGNA)	V1.4.1; 2015-02
EN 300 392-12-23	SS Stage 3: Call completion on no reply (CCNR)	V1.2.1; 2012-04
ETS 300 392-12-24	SS Stage 3: Call retention (CRT)	Edition 1; Sep 00
ETS 300 392-13	SDL Model of Air Interface (out of date and not maintained)	Edition 1; May 97 (historical)
392-16	see TS page for V+D Part 16	
392-17	see TR page for V+D Part 17	
392-18	see TS page for V+D Part 18	

TETRA Conformance testing specification		
EN 300 394-1	Conformance Testing Specification - Radio	V3.3.1; 2015-04

TETRA Speech Codec for full-rate traffic channel		
EN 300 395-1	General Description of Speech Functions	V1.2.1; 2005-01
EN 300 395-2	TETRA Codec	V1.3.1; 2005-01
EN 300 395-3	Specific Operating Features	V1.2.1; 2005-01
EN 300 395-4	CODEC Conformance Testing	V1.3.1; 2005-06

(Continued)

Technical Requirements for Direct Mode Operation (DMO)		
EN 300 396-1	DMO: General Network Design	V1.2.1; 2011-12
EN 300 396-2	DMO: Radio Aspects	V1.4.1; 2011-12
EN 300 396-3	DMO: Mobile Station to Mobile Station (MS-MS) Air Interface (AI) protocol	V1.4.1; 2011-12
EN 300 396-4	DMO: Type 1 Repeater Air Interface	V1.4.1; 2011-012
EN 300 396-5	DMO: Gateway Air Interface	V1.3.1; 2011-12
EN 300 396-6	DMO: Security	V1.6.1; 2016-11
ETS 300 396-8	**DMO: PICS proforma specifications**	

Subscriber Identity Module to Mobile Equipment (SIM - ME) interface		
EN 300 812-3	Subscriber Identity Module to Mobile Equipment (SIM - ME) interface; Part 3: Integrated Circuit (IC); Physical, logical and TSIM application characteristics	V2.3.1; 2005-12

TETRA Lawful Interception interface, EMC standards, End-to-end encryption		
EN 301 040	TETRA Security; Lawful Interception (LI) interface Note: see also associated ES 101 671 "Handover interface for the lawful interception of telecommunications traffic"	V2.1.1; 2006-03
EN 301 489-1	EMC standard for radio equipment and services: Common technical requirements	V2.1.1; 2017-02
EN 301 489-5	EMC standard for radio equipment and services: Specific conditions for Private land Mobile Radio (PMR) and ancillary equipment (speech and non-speech) and Terrestrial Trunked Radio (TETRA); Harmonised Standard covering the essential requirements of article 3.1(b) of the Directive 2014/53/EU (Note: includes information on TETRA previously in EN 301 489-18 which was not included prior to V2.1.0)	V2.1.1; 2016-11
EN 302 109	TETRA Security; Synchronisation mechanism for end-to-end encryption	V1.1.1; 2003-10

EN 302 561	ERM; Land Mobile Service; Radio equipment using constant or non-constant envelope modulation operating in a channel bandwidth of 25 kHz, 50 kHz, 100 kHz or 150 kHz; Harmonized EN covering essential requirements of article 3.2 of the R&TTE Directive	V2.1.1; 2016-03
EN 303 035-1	Harmonized EN for TETRA equipment covering essential requirements under article 3.2 of the R&TTE directive; Part 1: Voice plus Data (V+D)	V1.2.1; 2001-12
EN 303 035-2	Harmonized EN for TETRA equipment covering essential requirements under article 3.2 of the R&TTE directive; Part 2: Direct Mode Operation (DMO)	V1.2.2; 2003-01
EN 303 039	Multichannel transmitter specification for the PMR Service; Harmonized EN covering the essential requirements of article 3.2 of the R&TTE Directive	V2.1.2; 2016-10

B.2 ETSI STANDARDS (ESs)

SIM-ME, Lawful Interception interface, TAPS, End-to-end encryption		
ES 200 812-1	Subscriber Identity Module to Mobile Equipment (SIM - ME) interface; Part 1: Universal Integrated Circuit Card (UICC); Physical and logical characteristics Note - also published as TS 100 812-1 V2.2.5	V2.2.5; 2003-12
ES 200 812-2	Subscriber Identity Module to Mobile Equipment (TSIM - ME) interface; Part 2: Universal Integrated Circuit Card (UICC); Characteristics of the TSIM Application Note - also published as TS 100 812-2 V2.4.1	V2.4.2; 2005-10
	Note: see EN 300 812-3 for Part 3 of the SIM-ME Interface	
ES 201 671	Lawful Interception (LI); Handover interface for the lawful interception of telecommunications traffic Note 1: TS 101 671 V3.8.1 2011-08 presumably supersedes this ES Note 2: this is not directly related to TETRA but is referred to in Annex E of EN 301 040 which describes a Lawful Interception interface in a TETRA system	V3.1.1; 2007-05

(Continued)

| ES 201 962 | TETRA Advanced Packet Service (TAPS) | V1.1.1; 2001-09 |
| ES 202 109 | Security; Synchronisation mechanism for end-to-end encryption | V1.1.1; 2003-01 |

B.3 TECHNICAL SPECIFICATIONS (TSs)

	Interworking at ISI (speech format), V+D Security, Frequency bands & channel spacing, Network performance metrics, Location Information Protocol, Net Assist Protocol, Encryption algorithms, LI interworking, IP interworking, TMO Repeaters, RFSA-Tx Inhibit, TAPS test purposes, QoS aspects for popular services in mobile networks, Critical Communications Architecture	
TS 100 392-2	Voice + Data Air Interface (AI) Note: this has been superseded by published EN 300 392-2 V3.8.1 2016-08	V3.7.1; 2016-01
TS 100 392-3-8	Interworking at the Inter-System Interface (ISI); Generic Speech Format Implementation	V1.2.1; 2010-03
392-3-1, -2, -3, -4, -5	Note - see EN 300 392-3-1, -2, -3, -4, -5 for previous parts of ISI interworking	
TS 100 392-5	Peripheral Equipment Interface Note: this has been superseded by published EN 300 392-5 V2.5.1 2016-10	V2.4.1; 2016-03
TS 100 392-15	TETRA frequency bands, duplex spacing and channel numbering	V1.5.1; 2011-02
TS 100 392-16	Voice plus Data (V+D); Network Performance Metrics	V1.2.1; 2006-09
TS 100 392-18	**Voice plus Data (V+D); Air interface optimised applications**	
TS 100 392-18-1	Voice plus Data (V+D) Air interface optimised applications; Location Information Protocol (LIP)	V1.7.1; 2015-03
TS 100 392-18-2	Voice plus Data (V+D) Air interface optimized applications; Net Assist Protocol (NAP)	V1.1.1; 2008-11
TS 100 392-18-3	Voice plus Data (V+D) Air interface optimized applications; Direct mode Over The Air Management protocol (DOTAM)	V1.2.1; 2010-12
TS 100 392-18-4	Voice plus Data (V+D) Air interface optimized applications; Net Assist Protocol 2 (NAP2)	V1.2.1; 2015-07
TS 100 812-1	Subscriber Identity Module to Mobile Equipment (SIM - ME) interface; Part 1: Universal Integrated Circuit Card (UICC); Physical and logical characteristics Note: see ES 200 812-1 V2.2.5 for identical ES version	V2.2.5; 2003-10

TS 100 812-2	Subscriber Identity Module to Mobile Equipment (TSIM - ME) interface; Part 2: Universal Integrated Circuit Card (UICC); Characteristics of the TSIM Application Note: see ES 200 812-2 V2.4.2 for identical ES version	V2.4.1; 2005-08
	Note: see EN 300 812-3 for Part 3 of the SIM-ME Interface	
TS 101 052	Rules for the management of the TETRA standard authentication and key management algorithm set TAA1 Note: Previously published as TR 101 052	V2.1.1; 2016-02
TS 101 053-1	Rules for the management of the TETRA standard encryption algorithms; Part 1: TEA1 Note: Previously published as TR 101 053-1	V2.1.1; 2016-02
TS 101 053-2	Rules for the management of the TETRA standard encryption algorithms; Part 2: TEA2 Note: Previous published as TR 101 053-2	V2.3.1; 2014-04
TS 101 053-3	Rules for the management of the TETRA standard encryption algorithms; Part 3: TEA3 Note: Previously published as TR 101 053-3	V2.1.1; 2016-02
TS 101 053-4	Rules for the management of the TETRA standard encryption algorithms; Part 4: TEA4 Note: Previously published as TR 101 053-4	V2.1.1; 2016-02
TS 101 671	Lawful Interception (LI); Handover interface for the lawful interception of telecommunications traffic Note 1: assume this supersedes ES 201 671 V3.3.1 2007-05 Note 2: this is not directly related to TETRA but is referred to in Annex E of EN 301 040 which describes a Lawful Interception interface in a TETRA system	V3.14.1; 2016-03
TS 101 747	TETRA V+D; IP Interworking (IPI)	V1.1.1; 2001-07
TS 101 789-1	TETRA; TMO Repeaters Part 1: Requirements, test methods, and limits	V1.1.2; 2007-04
TS 101 975	RF Sensitive Area Mode Note: previously published version was TR 101 975 V1.1.1 now superseded	V1.2.1; 2007-07

(Continued)

TS 102 250	**STQ; QoS aspects for popular services in mobile networks** **Note: now contains some Quality of Service measurements applicable to TETRA**	
TS 102 250-1	STQ; QoS aspects for popular services in mobile networks; Part 1: Assessment of Quality of Service Note: Current version now includes TETRA-related general aspects by the removal of previous GSM and UMTS specific content	V2.2.1; 2011-04
TS 102 250-2	STQ; QoS aspects for popular services in mobile networks; Part 2: Definition of Quality of Service parameters and their computation Note: Current version includes additional QoS parameter definitions for TETRA individual call (§6.6), group call (§6.14), SDS (§7.4) & packet data services (§§5.5 & 5.6)	V2.5.1; 2016-06
TS 102 250-3	STQ; QoS aspects for popular services in mobile networks; Part 3: Typical procedures for Quality of Service measurement equipment Note: Current version includes new measurement procedures regarding TETRA service measurement procedures, especially for SDS (§8.5)	V2.3.2; 2015-08
TS 102 250-4	STQ; QoS aspects for popular services in mobile networks; Part 4: Requirements for Quality of Service measurement equipment Note: Current version includes TETRA specific requirements for measurement equipment, i.e. standardized PEI support (§5.4)	V2.2.1; 2011-04
TS 102 250-5	STQ; QoS aspects for popular services in mobile networks; Part 5: Definition of typical measurement profiles Note: Current version includes TETRA specific measurement profiles for individual call (§4.2.1.1), SDS measurements (§4.2.2.1) and packet data services (§4.2.3.2.1)	V2.4.2; 2015-09
TS 102 250-6	STQ; QoS aspects for popular services in GSM and 3G networks; Part 6: Post processing and statistical methods	Revision applicable to TETRA nya

TS 102 250-7	STQ; QoS aspects for popular services in GSM and 3G networks; Part 7: Network based Quality of Service measurements	Revision applicable to TETRA nya
TS 103 269-2	Critical Communications Architecture; Part 2: Critical Communications application mobile to network interface architecture; Note: see TR 103 269-1 for Part 1	V1.1.1; 2015-01

B.4 TECHNICAL REPORTS (TRs AND ETRs)

	TETRA V+D and DMO Release specifications, Encryption & authentication algorithms (TR series), ISI for 3CP, TRS on M-DMO, Guide to TAPS, AMR codec study, Air Interface Enhancements feasibility assessment, TAPS SRDoc, User Requirement Specifications, Designers' Guides, TEDS SRDoc, Security requirements for modulation extensions, TEDS using "Tuning Range" concept in 410-430 MHz and 450-470 MHz bands, Evaluation of low-rate codec, Future TETRA workshop report, Critical Communications Architecture (TR)	
TR 100 392-17-1	TETRA V+D and DMO specifications; Release 1.1 Note: previously published as TS 100 392-17 V1.1.1	V1.1.3; 2005-02
TR 100 392-17-2	TETRA V+D and DMO specifications; Release 1.2 Note: supersedes TS 100 392-17 Draft V1.1.8 due to the decision to publish as a TR	V1.1.1; 2004-06
TR 100 392-17-3	TETRA V+D and DMO specifications; Release 1.3	V1.2.1; 2006-06
TR 100 392-17-4	TETRA V+D and DMO specifications; Release 2.0	V1.1.1; 2008-05
TR 100 392-17-5	TETRA V+D and DMO specifications; Release 2.1	V1.1.1; 2011-11
TR 101 052	Rules for the management of the TETRA standard authentication and key management algorithm set TAA1 Note: Now published as TS 101 052 - see TS page	See TS 101 052 V2.1.1 2016-02
TR 101 053-1	Rules for the management of the TETRA standard encryption algorithms; Part 1: TEA1 Note: Now published as TS 101 053-1 - see TS page	See TS 101 053-1 V2.1.1 2016-02

(Continued)

TR 101 053-2	Rules for the management of the TETRA standard encryption algorithms; Part 2: TEA2 Note: Now published as TS 101 053-2 - see TS page	See TS 101 053-2 V2.3.1 2014-04
TR 101 053-3	Rules for the management of the TETRA standard encryption algorithms; Part 3: TEA3 Note: Now published as TS 101 053-3 - see TS page	See TS 101 053-3 V2.1.1 2016-02
TR 101 053-4	Rules for the management of the TETRA standard encryption algorithms; Part 4: TEA4 Note: Now published as TS 101 053-4 - see TS page	See TS 101 053-4 V2.1.1 2016-02
TR 101 448	Functional requirements for the TETRA ISI derived from the Three-Country Pilot Scenarios	V1.1.1; 2005-05
TR 101 977	Study of the suitability of the GSM Adaptive Multi-Rate (AMR) speech codec for use in TETRA	V1.1.1; 2001-07
TR 101 987	Proposed Air Interface Enhancements for TETRA Release 2; Analysis and Feasibility Assessment	V1.1.1; 2001-08
TR 102 001	Systems reference document for TETRA Advanced Packet Service (TAPS)	V1.1.1; 2003-04
TR 102 021-1	User Requirement Specification TETRA Release 2.1; Part 1: General Overview	V1.3.1; 2011-07
TR 102 021-2	User Requirement Specification TETRA Release 2.1; Part 2: High Speed Data	V1.3.1; 2010-12
TR 102 021-3	User Requirement Specification TETRA Release 2; Part 3: Codec	V1.1.1; 2001-12
TR 102 021-4	User Requirement Specification TETRA Release 2.1; Part 4: Air Interface Enhancements	V1.4.1; 2011-08
TR 102 021-5	User Requirement Specification TETRA Release 2.1; Part 5: Interworking and Roaming	V1.2.1; 2010-12

TR 102 021-6	User Requirement Specification TETRA Release 2.1; Part 6: Smart Card (SC) and Subscriber Identity Module (SIM)	V1.2.1; 2011-08
TR 102 021-7	User Requirement Specification TETRA Release 2.1; Part 7: Security	V1.3.1; 2010-12
TR 102 021-8	User Requirement Specification TETRA Release 2; Part 8: Air - Ground - Air services	V1.1.1; 2003-09
TR 102 021-9	User Requirement Specification TETRA Release 2.1; Part 9: Peripheral Equipment Interface	V1.2.1; 2010-12
TR 102 021-10	User Requirement Specification TETRA Release 2.1; Part 10: Local Mode Broadband	V1.1.1; 2010-12
TR 102 021-11	User Requirement Specification TETRA Release 2.1; Part 11: Over The Air Management	V1.1.1; 2011-07
TR 102 021-12	User Requirement Specification TETRA Release 2.1; Part 12: Direct Mode Operation (DMO)	V1.1.1; 2012-08
TR 102 022-1	User Requirement Specification; Mission Critical Broadband Communications [Part 1] Requirements	V1.1.1: 2012-08
TR 102 022-2	User Requirement Specification; Mission Critical Broadband Communications; Part 2: Critical Communications Application	V1.1.1: 2015-01
TR 102 300-2	Designers' guide; Part 2: Radio channels, network protocols and service performance Note: see ETR 300 series for Designers' Guides parts 1 & 4 and TR 102 580 for TEDS	V1.2.1; 2013-09
TR 102 300-3	Designers' Guide; Part 3: Direct Mode Operation (DMO) Note: see ETR 300 series for Designers' Guides parts 1 & 4 and TR 102 580 (below) for TEDS	V1.3.3; 2009-06

(*Continued*)

TR 102 300-5	Designers' Guide; Part 5: Guidance on numbering and addressing Note: see ETR 300 series for Designers' Guides parts 1 & 4 and TR 102 580 (below) for TEDS	V1.4.1; 2015-06
TR 102 300-6	Designers' Guide; Part 6: Air-Ground-Air Note: see ETR 300 series for Designers' Guides parts 1 & 4 and TR 102 580 (below) for TEDS	V1.1.2; 2016-05
TR 102 300-7	Designers' Guide; Part 7: TETRA High-Speed Data (HSD); TETRA Enhanced Data Service (TEDS) Note 1: previously published as TR 102 580 Note 2: see ETR 300 series for Designers' Guides parts 1 & 4	V1.2.1; 2016-11
TR 102 459	TETRA Air-Ground-Air services (AGA); System reference document	V1.1.1; 2006-05
TR 102 491	TETRA Enhanced Data Service (TEDS); System reference document	V1.2.1; 2006-05
TR 102 512	Security requirements analysis for modulation extensions to TETRA	V1.1.1; 2006-08
TR 102 513	Feasibility Study into the Implications of Operating Public Safety Sector (PSS) TEDS using the proposed "Tuning Range" concept in the 410-430 MHz and 450-470 MHz frequency bands	V1.1.1; 2006-12
TR 102 582	Evaluation of low rate (2,4 kbit/s) speech codec	V1.1.1; 2007-07
TR 102 621	TWC2007 Future of TETRA workshop report	V1.1.1; 2008-04
TR 102 628	ERM: System reference document; Land Mobile Service; Additional spectrum requirements for future Public Safety and Security (PSS) wireless communication systems in the UHF frequency range	V1.2.1; 2014-09
	Associated LEWP matrix (see Annex F.6 of TR 102 628)	
TR 102 753	TETRA mobiles moving at high velocity	V1.1.1; 2008-05

TR 103 269-1	Critical Communications Architecture; Part 1: Critical Communications Architecture Reference Model Note: see TR 103 269-2 for Part 2	V1.1.1; 2014-07
TR 103 414	Study into the provision of speech services over QAM channels	V1.1.1; 2016-09

B.5 ETSI GUIDES (EGs)

Radio site engineering, TETRA numbering and its administration		
EG 200 053	Radio site engineering for radio equipment and systems	V1.5.1; 2004-06
EG 202 118	The structure of the TETRA numbering resource, interworking and high level policy for administration	V1.1.1; 2003-05

C

LTE CRITICAL COMMUNICATIONS RELATED DOCUMENTS

This appendix provides a list of 3GPP technical standards and reports specific only to the critical communications and public safety applications. For a full list of 3GPP documents see (provide a site where we can place the complete LTE standards.)

Source: http://www.3gpp.org/DynaReport/status-report.htm#activeRel-16

C.1 CRITICAL COMMUNICATIONS RELATED 3GPP TECHNICAL STANDARDS (TSs)

Number	Title	Rel. 8	Rel. 9	Rel. 10	Rel. 11	Rel. 12	Rel. 13	Rel. 14
TS 22.179	Mission Critical Push to Talk (MCPTT) over LTE; Stage 1						13.3.0	14.3.0
TS 22.268	Public Warning System (PWS) requirements		9.5.0		11.5.0	12.2.0	13.0.0	14.1.0

Fundamentals of Public Safety Networks and Critical Communications Systems: Technologies, Deployment, and Management, First Edition. Mehmet Ulema.
© 2019 by The Institute of Electrical and Electronics Engineers, Inc. Published 2019 by John Wiley & Sons, Inc.

Number		Title	Rel. 8	Rel. 9	Rel. 10	Rel. 11	Rel. 12	Rel. 13	Rel. 14
TS	22.280	Mission Critical Services Common Requirements							14.3.0
TS	22.281	Mission Critical Video over LTE							14.3.0
TS	22.282	Mission Critical Data over LTE							14.3.0
TS	22.346	Isolated Evolved Universal Terrestrial Radio Access Network (E-UTRAN) operation for public safety; Stage 1						13.0.0	14.0.0
TS	23.179	Functional architecture and information flows to support mission critical communication services; Stage 2						13.5.0	
TS	23.280	Common functional architecture to support mission critical services; Stage 2							14.3.0
TS	23.281	Functional architecture and information flows to support Mission Critical Video (MCVideo); Stage 2							14.3.0
TS	23.282	Functional architecture and information flows to support Mission Critical Data (MCData); Stage 2							14.3.0
TS	23.379	Functional architecture and information flows to support Mission Critical Push To Talk (MCPTT); Stage 2							14.3.0

Number	Title	Rel. 8	Rel. 9	Rel. 10	Rel. 11	Rel. 12	Rel. 13	Rel. 14
TS 23.509	TISPAN; NGN Architecture to support emergency communication from citizen to authority [Endorsed document 3GPP TS 23.167, Release 7]	8.0.0						
TS 24.281	Mission Critical Video (MCVideo) signalling control; Protocol specification							14.1.0
TS 24.282	Mission Critical Data (MCData) signalling control; Protocol specification							14.1.0
TS 24.379	Mission Critical Push To Talk (MCPTT) call control; Protocol specification						13.6.0	14.3.0
TS 24.380	Mission Critical Push To Talk (MCPTT) media plane control; Protocol specification						13.6.0	14.4.0
TS 24.381	Mission Critical Push To Talk (MCPTT) group management; Protocol specification						13.4.0	
TS 24.382	Mission Critical Push To Talk (MCPTT) identity management; Protocol specification						13.3.0	
TS 24.383	Mission Critical Push To Talk (MCPTT) Management Object (MO)						13.4.0	
TS 24.384	Mission Critical Push To Talk (MCPTT) configuration management; Protocol specification						13.4.0	

(Continued)

Number		Title	Rel. 8	Rel. 9	Rel. 10	Rel. 11	Rel. 12	Rel. 13	Rel. 14
TS	24.481	Mission Critical Services (MCS) group management; Protocol specification						13.6.0	14.2.0
TS	24.482	Mission Critical Services (MCS) identity management; Protocol specification						13.3.0	14.2.0
TS	24.483	Mission Critical Services (MCS) Management Object (MO)						13.6.0	14.2.0
TS	24.484	Mission Critical Services (MCS) configuration management; Protocol specification						13.6.0	14.3.1
TS	24.581	Mission Critical Video (MCVideo) media plane control; Protocol specification							14.1.0
TS	24.582	Mission Critical Data (MCData) media plane control; Protocol specification							14.1.0
TS	26.179	Mission Critical Push To Talk (MCPTT); Codecs and media handling						13.2.0	14.0.0
TS	26.281	Mission Critical Video (MCVideo); Codecs and media handling							14.0.0
TS	36.579-5	Mission Critical Push To Talk (MCPTT) over LTE; Part 5: Abstract test suite (ATS)						0.0.1	none

Number	Title	Rel. 8	Rel. 9	Rel. 10	Rel. 11	Rel. 12	Rel. 13	Rel. 14
TS 36.579-4	Mission Critical Push To Talk (MCPTT) over LTE; Part 4: Test Applicability and Implementation Conformance Statement (ICS) proforma specification						0.0.1	0.2.0
TS 33.179	Security of Mission Critical Push To Talk (MCPTT) over LTE						13.4.0	
TS 33.180	Security of the mission critical service							14.1.0
TS 33.269	Public Warning System (PWS) security architecture					none	none	none
TS 33.897	Study on isolated E-UTRAN operation for public safety; Security aspects						13.1.0	
TS 36.579-3	Mission Critical Push To Talk (MCPTT) over LTE; Part 3: MCPTT Server Application conformance specification							none
TS 36.579-2	Mission Critical Push To Talk (MCPTT) over LTE; Part 2: User Equipment (UE) Protocol conformance specification							14.3.0
TS 36.579-2	Mission Critical Push To Talk (MCPTT) over LTE; Part 2: User Equipment (UE) Protocol conformance specification						13.0.1	
TS 36.579-1	Mission Critical Push To Talk (MCPTT) over LTE; Part 1: Common test environment							14.0.1

C.2 CRITICAL COMMUNICATIONS RELATED 3GPP TECHNICAL REPORTS (TRs)

Number	Title	Rel. 8	Rel. 9	Rel. 10	Rel. 11	Rel. 12	Rel. 13	Rel. 14
TR 22.815	Study on Multimedia Broadcast Supplement for Public Warning System (MBSP)							14.0.0
TR 22.879	Feasibility study on mission critical video services over LTE							14.0.0
TR 22.880	Feasibility study on mission critical data communications							14.0.0
TR 22.897	Study on isolated Evolved Universal Terrestrial Radio Access Network (E-UTRAN) operation for public safety						13.0.0	
TR 22.968	Study for requirements for a Public Warning System (PWS) service	8.0.0	9.0.0	10.0.0	11.0.0	12.0.0	13.0.0	14.0.0
TR 23.779	Study on application architecture to support Mission Critical Push To Talk over LTE (MCPTT) services						13.0.0	
TR 23.780	Study on Multimedia Broadcast and Multicast Service (MBMS) usage for mission critical communication services							14.0.0
TR 23.781	Study on migration and interconnection for mission critical services							14.0.0
TR 23.797	Study on architecture enhancements to support isolated Evolved Universal Terrestrial Radio Access Network (E-UTRAN) operation for public safety						13.0.0	

Number	Title	Rel. 8	Rel. 9	Rel. 10	Rel. 11	Rel. 12	Rel. 13	Rel. 14
TR 23.828	Earthquake and Tsunami Warning System (ETWS); Requirements and solutions; Solution placeholder	8.0.0						
TR 24.980	Minimum requirements for support of Mission Critical Push To Talk (MCPTT) service over the Gm reference point						13.3.0	14.1.0
TR 25.703	Universal Terrestrial Radio Access (E-UTRA); Study on Home Node B (HNB) emergency warning area					12.0.0		
TR 26.989	Mission Critical Push To Talk (MCPTT); Media, codecs and Multimedia Broadcast/Multicast Service (MBMS) enhancements for MCPTT over LTE						13.1.0	14.0.0
TR 32.844	Study of charging support of Proximity-based Services (ProSe) direct communication for public safety use					12.0.0		
TR 33.879	Study on security enhancements for Mission Critical Push To Talk (MCPTT) over LTE						13.1.0	
TR 33.969	Study on security aspects of Public Warning System (PWS)					12.0.0	13.0.0	14.0.0
TR 36.837	Public safety broadband high power User Equipment (UE) for band 14				11.0.0			

INDEX

Fundamentals of Public Safety Networks and Critical Communications Systems: Technologies, Deployment, and Management,
First Edition. Mehmet Ulema.
© 2019 by The Institute of Electrical and Electronics Engineers, Inc. Published 2019 by John Wiley & Sons, Inc.

IEEE Press Series on
Networks and Services Management

The goal of this series is to publish high quality technical reference books and textbooks on network and services management for communications and information technology professional societies, private sector and government organizations as well as research centers and universities around the world. This Series focuses on Fault, Configuration, Accounting, Performance, and Security (FCAPS) management in areas including, but not limited to, telecommunications network and services, technologies and implementations, IP networks and services, and wireless networks and services.

Series Editors:
Dr. Veli Sahin
Dr. Mehmet Ulema

Fundamentals of Public Safety Networks and Critical Communications Systems: Technologies, Deployment, and Management, First Edition. Mehmet Ulema.

© 2019 by The Institute of Electrical and Electronics Engineers, Inc. Published 2019 by John Wiley & Sons, Inc.